Six Easy
Pieces

SIX EASY PIECES

Essentials of Physics
Explained by Its Most
Brilliant Teacher

RICHARD P. FEYNMAN

Originally prepared for publication by
Robert B. Leighton and Matthew Sands

New Introduction by Paul Davies

HELIX BOOKS

PERSEUS BOOKS
Cambridge, Massachusetts

Many of the designations used by manufacturers and sellers to distinguish their products are claimed as trademarks. Where those designations appear in this book and Perseus Books was aware of a trademark claim, the designations have been printed in initial capital letters.

Library of Congress Cataloging-in-Publication Data

Feynman, Richard Phillips.
   Six easy pieces : essentials of physics explained by its most
 brilliant teacher / Richard P. Feynman ; originally prepared for
 publication by Robert B. Leighton and Matthew Sands.
     p. cm.
   Includes index.
   ISBN 0-201-40955-0 (hardcover), 0-201-40825-2 (paperback)
   1. Physics.   I. Leighton, Robert B.   II. Sands, Matthew L.
(Matthew Linzee)   III. Title.   IV. Title: 6 easy pieces.
QC21.2.F52   1994
530—dc20   94-30894
                                          CIP

Perseus Books is a member of the Perseus Books Group

Cover and text design by Diane Levy
Set in 11-point Simoncini Garamond by Pagesetters, Inc.

     17 18 19 20 –DOH–02

Find us on the World Wide Web at
http://www.aw.com/gb/

Contents

Contents

Publisher's Note

SIX EASY PIECES grew out of the need to bring to as wide an audience as possible a substantial yet non-technical physics primer based on the science of Richard Feynman. We have chosen the six easiest chapters from Feynman's celebrated and landmark text, Lectures on Physics (originally published in 1963), which remains his most famous publication. General readers are fortunate that Feynman chose to present certain key topics in largely qualitative terms without formal mathematics, and these are brought together for SIX EASY PIECES.

Addison-Wesley Publishing Company would like to thank Paul Davies for his insightful introduction to this newly formed collection. Following his introduction we have chosen to reproduce two prefaces from Lectures on Physics, one by Feynman himself and one by two of his colleagues, because they provide context for the pieces that follow and insight into both Richard Feynman and his science.

Finally, we would like to thank the California Institute of Technology's Physics Department and Institute Archives, in particular Dr. Judith Goodstein, and Dr. Brian Hatfield, for his outstanding advice and recommendations throughout the development of this project.

Introduction

There is a popular misconception that science is an impersonal, dispassionate, and thoroughly objective enterprise. Whereas most other human activities are dominated by fashions, fads, and personalities, science is supposed to be constrained by agreed rules of procedure and rigorous tests. It is the results that count, not the people who produce them.

This is, of course, manifest nonsense. Science is a people-driven activity like all human endeavor, and just as subject to fashion and whim. In this case fashion is set not so much by choice of subject matter, but by the way scientists think about the world. Each age adopts its particular approach to scientific problems, usually following the trail blazed by certain dominant figures who both set the agenda and define the best methods to tackle it. Occasionally scientists attain sufficient stature that they become noticed by the general public, and when endowed with outstanding flair a scientist may become an icon for the entire scientific community. In earlier centuries Isaac Newton was an icon. Newton personified the gentleman scientist—well-connected, devoutly religious, unhurried, and methodical in his work. His style of doing science set the standard for two hundred years. In the first half of the twentieth century Albert Einstein replaced Newton as the popular scientist icon. Eccentric, dishevelled, Germanic, absent-minded, utterly absorbed in his work, and an archetypal abstract thinker, Einstein changed the way that physics is done by questioning the very concepts that define the subject.

Richard Feynman has become an icon for late twentieth-century physics—the first American to achieve this status. Born in New York in 1918 and educated on the East Coast, he was too late to participate in the Golden Age of physics, which, in the first three decades of this century, transformed our world view with the twin revolutions of the theory of relativity and quantum mechanics. These sweeping developments laid the foundations of the edifice we now call the New Physics. Feynman started with those foundations and helped build the ground floor of the New Physics. His contributions touched almost every corner of the subject and have had a deep and abiding influence over the way that physicists think about the physical universe.

Feynman was a theoretical physicist par excellence. Newton had been both experimentalist and theorist in equal measure. Einstein was quite simply contemptuous of experiment, preferring to put his faith in pure thought. Feynman was driven to develop a deep theoretical understanding of nature, but he always remained close to the real and often grubby world of experimental results. Nobody who watched the elderly Feynman elucidate the cause of the Challenger space shuttle disaster by dipping an elastic band in ice water could doubt that here was both a showman and a very practical thinker.

Initially, Feynman made a name for himself from his work on the theory of subatomic particles, specifically the topic known as quantum electrodynamics or QED. In fact, the quantum theory began with this topic. In 1900, the German physicist Max Planck proposed that light and other electromagnetic radiation, which had hitherto been regarded as waves, paradoxically behaved like tiny packets of energy, or "quanta," when interacting with matter. These particular quanta became known as photons. By the early 1930s the architects of the new quantum mechanics had worked out a mathematical scheme to describe the emission and absorption of photons by electrically charged particles such as electrons. Although this early formulation of QED enjoyed some limited success, the theory was clearly flawed. In many cases calculations gave inconsistent and even infinite answers to well-posed physical questions. It was to the problem of constructing a consistent theory of QED that the young Feynman turned his attention in the late 1940s.

To place QED on a sound basis it was necessary to make the theory consistent not only with the principles of quantum mechanics but with those of the special theory of relativity too. These two theories come with their own distinctive mathematical machinery, complicated systems of equations that can indeed be combined and reconciled to yield a satisfactory description of QED. Doing this was a tough undertaking, requiring a high degree of mathematical skill, and was the approach followed by Feynman's contemporaries. Feynman himself, however, took a radically different route—so radical, in fact, that he was more or less able to write down the answers straight away without using any mathematics!

To aid this extraordinary feat of intuition, Feynman invented a simple system of eponymous diagrams. Feynman diagrams are a symbolic but powerfully heuristic way of picturing what is going on when electrons, photons, and other particles interact with each other. These days Feynman diagrams are a routine aid to calculation, but in the early 1950s they marked a startling departure from the traditional way of doing theoretical physics.

The particular problem of constructing a consistent theory of quantum electrodynamics, although it was a milestone in the development of physics, was just the start. It was to define a distinctive Feynman style, a style destined to produce a string of important results from a broad range of topics in physical science. The Feynman style can best be described as a mixture of reverence and disrespect for received wisdom.

Physics is an exact science, and the existing body of knowledge, while incomplete, can't simply be shrugged aside. Feynman acquired a formidable grasp of the accepted principles of physics at a very young age, and he chose to work almost entirely on conventional problems. He was not the sort of genius to beaver away in isolation in a backwater of the discipline and to stumble across the profoundly new. His special talent was to approach essentially mainstream topics in an idiosyncratic way. This meant eschewing

Introduction

existing formalisms and developing his own highly intuitive approach. Whereas most theoretical physicists rely on careful mathematical calculation to provide a guide and a crutch to take them into unfamiliar territory, Feynman's attitude was almost cavalier. You get the impression that he could read nature like a book and simply report on what he found, without the tedium of complex analysis.

Indeed, in pursuing his interests in this manner Feynman displayed a healthy contempt for rigorous formalisms. It is hard to convey the depth of genius that is necessary to work like this. Theoretical physics is one of the toughest intellectual exercises, combining abstract concepts that defy visualization with extreme mathematical complexity. Only by adopting the highest standards of mental discipline can most physicists make progress. Yet Feynman appeared to ride roughshod over this strict code of practice and pluck new results like ready-made fruit from the Tree of Knowledge.

The Feynman style owed a great deal to the personality of the man. In his professional and private life he seemed to treat the world as a hugely entertaining game. The physical universe presented him with a fascinating series of puzzles and challenges, and so did his social environment. A lifelong prankster, he treated authority and the academic establishment with the same sort of disrespect he showed for stuffy mathematical formalism. Never one to suffer fools gladly, he broke the rules whenever he found them arbitrary or absurd. His autographical writings contain amusing stories of Feynman outwitting the atom-bomb security services during the war, Feynman cracking safes, Feynman disarming women with outrageously bold behavior. He treated his Nobel Prize, awarded for his work on QED, in a similar take-it-or-leave-it manner.

Alongside this distaste for formality, Feynman had a fascination with the quirky and obscure. Many will remember his obsession with the long-lost country of Tuva in Central Asia, captured so delightfully in a documentary film made near to the time of his

Introduction

death. His other passions included playing the bongo drums, painting, frequenting strip clubs, and deciphering Mayan texts.

Feynman himself did much to cultivate his distinctive persona. Although reluctant to put pen to paper, he was voluble in conversation, and loved to tell stories about his ideas and escapades. These anecdotes, accumulated over the years, helped add to his mystique and made him a proverbial legend in his own lifetime. His engaging manner endeared him greatly to students, especially the younger ones, many of whom idolized him. When Feynman died of cancer in 1988 the students at Caltech, where he had worked for most of his career, unfurled a banner with the simple message: "We love you Dick."

It was Feynman's happy-go-lucky approach to life in general and physics in particular that made him such a superb communicator. He had little time for formal lecturing or even for supervising Ph.D. students. Nevertheless he could give brilliant lectures when it suited him, deploying all the sparkling wit, penetrating insight, and irreverence that he brought to bear on his research work.

In the early 1960s Feynman was persuaded to teach an introductory physics course to Caltech freshmen and sophomores. He did so with characteristic panache and his inimitable blend of informality, zest, and off-beat humor. Fortunately, these priceless lectures were saved for posterity in book form. Though far removed in style and presentation from more conventional teaching texts, the Feynman *Lectures on Physics* were a huge success, and they excited and inspired a generation of students across the world. Three decades on, these volumes have lost nothing of their sparkle and lucidity. *Six Easy Pieces* is culled directly from *Lectures on Physics*. It is intended to give general readers a substantive taste of Feynman the Educator by drawing on the early, non-technical chapters from that landmark work. The result is a delightful volume—it serves as both a primer on physics for non-scientists and as a primer on Feynman himself.

What is most impressive about Feynman's carefully crafted exposition is the way that he can develop far-reaching physical notions from the most slender investment in concepts, and a minimum in the way of mathematics and technical jargon. He has the knack of finding just the right analogy or everyday illustration to bring out the essence of a deep principle, without obscuring it in incidental or irrelevant details.

The selection of topics contained in this volume is not intended as a comprehensive survey of modern physics, but as a tantalizing taste of the Feynman approach. We soon discover how he can illuminate even mundane topics like force and motion with new insights. Key concepts are illustrated by examples drawn from daily life or antiquity. Physics is continually linked to other sciences while leaving the reader in no doubt about which is the fundamental discipline.

Right at the beginning of *SIX EASY PIECES* we learn how all physics is rooted in the notion of law—the existence of an ordered universe that can be understood by the application of rational reasoning. However, the laws of physics are not transparent to us in our direct observations of nature. They are frustratingly hidden, subtly encoded in the phenomena we study. The arcane procedures of the physicist—a mixture of carefully designed experimentation and mathematical theorizing—are needed to unveil the underlying lawlike reality.

Possibly the best-known law of physics is Newton's inverse square law of gravitation, discussed here in Chapter Five. The topic is introduced in the context of the solar system and Kepler's laws of planetary motion. But gravitation is universal, applying across the cosmos, enabling Feynman to spice his account with examples from astronomy and cosmology. Commenting on a picture of a globular cluster somehow held together by unseen forces, he waxes lyrical: "If one cannot see gravitation acting here, he has no soul."

Other laws are known that refer to the various non-gravitational forces of nature that describe how particles of matter interact with each other. There is but a handful of these forces, and Feynman himself holds the considerable distinction of being one of the few scientists in history to discover a new law of physics, pertaining to the way that a weak nuclear force affects the behavior of certain subatomic particles.

Introduction

High-energy particle physics was the jewel in the crown of postwar science, at once awesome and glamorous, with its huge accelerator machines and seemingly unending list of newly discovered subatomic particles. Feynman's research was directed mostly toward making sense of the results of this enterprise. A great unifying theme among particle physicists has been the role of symmetry and conservation laws in bringing order to the subatomic zoo.

As it happens, many of the symmetries known to particle physicists were familiar already in classical physics. Chief among these are the symmetries that arise from the homogeneity of space and time. Take time: apart from cosmology, where the big bang marked the beginning of time, there is nothing in physics to distinguish one moment of time from the next. Physicists say that the world is "invariant under time translations," meaning that whether you take midnight or mid-day to be the zero of time in your measurements, it makes no difference to the description of physical phenomena. Physical processes do not depend on an absolute zero of time. It turns out that this symmetry under time translation directly implies one of the most basic, and also most useful, laws of physics: the law of conservation of energy. This law says that you can move energy around and change its form but you can't create or destroy it. Feynman makes this law crystal clear with his amusing story of Dennis the Menace who is always mischievously hiding his toy building blocks from his mother (Chapter Four).

The most challenging lecture in this volume is the last, which is an exposition on quantum physics. It is no exaggeration to say that quantum mechanics had dominated twentieth-century physics and is far and away the most successful scientific theory in existence. It is indispensable for understanding subatomic particles, atoms and nuclei, molecules and chemical bonding, the structure of solids, superconductors and superfluids, the electrical and thermal conductivity of metals and semiconductors, the structure of stars, and much else. It has practical applications ranging from the laser to the microchip. All this from a theory that at first sight—and second sight—looks absolutely crazy! Niels Bohr, one of the founders of quantum mechanics, once remarked that anybody who is not shocked by the theory hasn't understood it.

The problem is that quantum ideas strike at the very heart of what we might call common-sense reality. In particular, the idea that physical objects such as electrons or atoms enjoy an independent existence, with a complete set of physical properties at all times, is called into question. For example, an electron cannot have a position in space and a well-defined speed at the same moment. If you look for where an electron is located, you will find it at a place, and if you measure its speed you will obtain a definite answer, but you cannot make both observations at once. Nor is it meaningful to attribute definite yet unknown values for the position and speed to an electron in the absence of a complete set of observations.

This indeterminism in the very nature of atomic particles is encapsulated by Heisenberg's celebrated uncertainty principle. This puts strict limits on the precision with which properties such as position and speed can be simultaneously known. A sharp value for position smears the range of possible values of speed and vice versa. Quantum fuzziness shows up in the way electrons, photons, and other particles move. Certain experiments can reveal them taking definite paths through space, after the fashion of bullets following trajectories toward a target. But other experimental arrangements reveal that these entities can also behave like waves, showing characteristic patterns of diffraction and interference.

Feynman's masterly analysis of the famous "two-slit" experiment, which teases out the "shocking" wave-particle duality in its starkest form, has become a classic in the history of scientific exposition. With a few very simple ideas, Feynman manages to take the reader to the very heart of the quantum mystery, and leaves us dazzled by the paradoxical nature of reality that it exposes.

Although quantum mechanics had made the textbooks by the early 1930s, it is typical of Feynman that, as a young man, he preferred to refashion the theory for himself in an entirely new guise. The Feynman method has the virtue that it provides us with a vivid picture of nature's quantum trickery at work. The idea is that

Introduction

the path of a particle through space is not generally well-defined in quantum mechanics. We can imagine a freely moving electron, say, not merely traveling in a straight line between A and B as common sense would suggest, but taking a variety of wiggly routes. Feynman invites us to imagine that somehow the electron explores all possible routes, and in the absence of an observation about which path is taken we must suppose that all these alternative paths somehow contribute to the reality. So when an electron arrives at a point in space—say a target screen—many different histories must be integrated together to create this one event.

Feynman's so-called path-integral, or sum-over-histories approach to quantum mechanics, set this remarkable concept out as a mathematical procedure. It remained more or less a curiosity for many years, but as physicists pushed quantum mechanics to its limits—applying it to gravitation and even cosmology—so the Feynman approach turned out to offer the best calculational tool for describing a quantum universe. History may well judge that, among his many outstanding contributions to physics, the path-integral formulation of quantum mechanics is the most significant.

Many of the ideas discussed in this volume are deeply philosophical. Yet Feynman had an abiding suspicion of philosophers. I once had occasion to tackle him about the nature of mathematics and the laws of physics, and whether abstract mathematical laws could be considered to enjoy an independent Platonic existence. He gave a spirited and skillful description of why this indeed appears so but soon backed off when I pressed him to take a specific philosophical position. He was similarly wary when I attempted to draw him out on the subject of reductionism. With hindsight, I believe that Feynman was not, after all, contemptuous of philosophical problems. But, just as he was able to do fine mathematical physics without systematic mathematics, so he produced some fine philosophical insights without systematic philosophy. It was formalism he disliked, not content.

It is unlikely that the world will see another Richard Feynman. He was very much a man of his time. The Feynman style worked

Introduction

well for a subject that was in the process of consolidating a revolution and embarking on the far-reaching exploration of its consequences. Postwar physics was secure in its foundations; mature in its theoretical structures, yet wide open for kibitzing exploitation. Feynman entered a wonderland of abstract concepts and imprinted his personal brand of thinking upon many of them. This book provides a unique glimpse into the mind of a remarkable human being.

September 1994   PAUL DAVIES

Special Preface

(from *Lectures on Physics*)

Toward the end of his life, Richard Feynman's fame had transcended the confines of the scientific community. His exploits as a member of the commission investigating the space shuttle Challenger disaster gave him widespread exposure; similarly, a best-selling book about his picaresque adventures made him a folk hero almost of the proportions of Albert Einstein. But back in 1961, even before his Nobel Prize increased his visibility to the general public, Feynman was more than merely famous among members of the scientific community—he was legendary. Undoubtedly, the extraordinary power of his teaching helped spread and enrich the legend of Richard Feynman.

He was a truly great teacher, perhaps the greatest of his era and ours. For Feynman, the lecture hall was a theater, and the lecturer a performer, responsible for providing drama and fireworks as well as facts and figures. He would prowl about the front of a classroom, arms waving, "the impossible combination of theoretical physicist and circus barker, all body motion and sound effects," wrote *The New York Times*. Whether he addressed an audience of students, colleagues, or the general public, for those lucky enough to see Feynman lecture in person, the experience was usually unconventional and always unforgettable, like the man himself.

He was the master of high drama, adept at riveting the attention of every lecture-hall audience. Many years ago, he taught a course in Advanced Quantum Mechanics, a large class comprised of a few registered graduate students and most of the Caltech physics faculty. During one lecture, Feynman started explaining how to represent certain complicated integrals diagrammatically: time on this axis, space on that axis, wiggly line for this straight line, etc. Having described what is known to the world of physics as a Feynman diagram, he turned around to face the class, grinning wickedly. "And this is called *THE* diagram!" Feynman had reached the denouement, and the lecture hall erupted with spontaneous applause.

For many years after the lectures that make up this book were given, Feynman was an occasional guest lecturer for Caltech's freshman physics course. Naturally, his appearances had to be kept secret so there would be room left in the hall for the registered students. At one such lecture the subject was curved-space time, and Feynman was characteristically brilliant. But the unforgettable moment came at the beginning of the lecture. The supernova of 1987 has just been discovered, and Feynman was very excited about it. He said, "Tycho Brahe had his supernova, and Kepler had his. Then there weren't any for 400 years. But now I have mine." The class fell silent, and Feynman continued on. "There are 10^{11} stars in the galaxy. That used to be a *huge* number. But it's only a hundred billion. It's less than the national deficit! We used to call them astronomical numbers. Now we should call them economical numbers." The class dissolved in laughter, and Feynman, having captured his audience, went on with his lecture.

Showmanship aside, Feynman's pedagogical technique was simple. A summation of his teaching philosophy was found among his papers in the Caltech archives, in a note he had scribbled to himself while in Brazil in 1952:

> "First figure out why you want the students to learn the subject and what you want them to know, and the method will result more or less by common sense."

What came to Feynman by "common sense" were often brilliant twists that perfectly captured the essence of his point. Once, during

Special Preface

a public lecture, he was trying to explain why one must not verify an idea using the same data that suggested the idea in the first place. Seeming to wander off the subject, Feynman began talking about license plates. "You know, the most amazing thing happened to me tonight. I was coming here, on the way to the lecture, and I came in through the parking lot. And you won't believe what happened. I saw a car with the license plate ARW 357. Can you imagine? Of all the millions of license plates in the state, what was the chance that I would see that particular one tonight? Amazing!" A point that even many scientists fail to grasp was made clear through Feynman's remarkable "common sense."

In 35 years at Caltech (from 1952 to 1987), Feynman was listed as teacher of record for 34 courses. Twenty-five of them were advanced graduate courses, strictly limited to graduate students, unless undergraduates asked permission to take them (they often did, and permission was nearly always granted). The rest were mainly introductory graduate courses. Only once did Feynman teach courses purely for undergraduates, and that was the celebrated occasion in the academic years 1961 to 1962 and 1962 to 1963, with a brief reprise in 1964, when he gave the lectures that were to become *The Feynman Lectures on Physics*.

At the time there was a consensus at Caltech that freshman and sophomore students were getting turned off rather than spurred on by their two years of compulsory physics. To remedy the situation, Feynman was asked to design a series of lectures to be given to the students over the course of two years, first to freshmen, and then to the same class as sophomores. When he agreed, it was immediately decided that the lectures should be transcribed for publication. That job turned out to be far more difficult than anyone had imagined. Turning out publishable books required a tremendous amount of work on the part of his colleagues, as well as Feynman himself, who did the final editing of every chapter.

And the nuts and bolts of running a course had to be addressed. This task was greatly complicated by the fact that Feynman had only a vague outline of what he wanted to cover. This meant that no one knew what Feynman would say until he stood in front of a lecture hall filled with students and said it. The Caltech professors who assisted him would then scramble as best they could to handle mundane details, such as making up homework problems.

Why did Feynman devote more than two years to revolutionize the way beginning physics was taught? One can only speculate, but there were probably three basic reasons. One is that he loved to have an audience, and this gave him a bigger theater than he usually had in graduate courses. The second was that he genuinely cared about students, and he simply thought that teaching freshmen was an important thing to do. The third and perhaps most important reason was the sheer challenge of reformulating physics, as he understood it, so that it could be presented to young students. This was his specialty, and was the standard by which he measured whether something was really understood. Feynman was once asked by a Caltech faculty member to explain why spin 1/2 particles obey Fermi-Dirac statistics. He gauged his audience perfectly and said, "I'll prepare a freshman lecture on it." But a few days later he returned and said, "You know, I couldn't do it. I couldn't reduce it to the freshman level. That means we really don't understand it."

This specialty of reducing deep ideas to simple, understandable terms is evident throughout *The Feynman Lectures on Physics*, but nowhere more so than in his treatment of quantum mechanics. To aficionados, what he has done is clear. He has presented, to beginning students, the path integral method, the technique of his own devising that allowed him to solve some of the most profound problems in physics. His own work using path integrals, among other achievements, led to the 1965 Nobel Prize that he shared with Julian Schwinger and Sin-Itero Tomanaga.

Through the distant veil of memory, many of the students and faculty attending the lectures have said that having two years of physics with Feynman was the experience of a lifetime. But that's not how it seemed at the time. Many of the students dreaded the class, and as the course wore on, attendance by the registered students started dropping alarmingly. But at the same time, more

Special Preface

and more faculty and graduate students started attending. The room stayed full, and Feynman may never have known he was losing some of his intended audience. But even in Feynman's view, his pedagogical endeavor did not succeed. He wrote in the 1963 preface to the *Lectures:* "I don't think I did very well by the students." Rereading the books, one sometimes seems to catch Feynman looking over his shoulder, not at his young audience, but directly at his colleagues, saying, "Look at that! Look how I finessed that point! Wasn't that clever?" But even when he thought he was explaining things lucidly to freshmen or sophomores, it was not really they who were able to benefit most from what he was doing. It was his peers—scientists, physicists, and professors—who would be the main beneficiaries of his magnificent achievement, which was nothing less than to see physics through the fresh and dynamic perspective of Richard Feynman.

Feynman was more than a great teacher. His gift was that he was an extraordinary teacher of teachers. If the purpose in giving *The Feynman Lectures on Physics* was to prepare a roomful of undergraduate students to solve examination problems in physics, he cannot be said to have succeeded particularly well. Moreover, if the intent was for the books to serve as introductory college textbooks, he cannot be said to have achieved his goal. Nevertheless, the books have been translated into ten foreign languages and are available in four bilingual editions. Feynman himself believed that his most important contribution to physics would not be QED, or the theory of superfluid helium, or polarons, or partons. His foremost contribution would be the three red books of *The Feynman Lectures on Physics*. That belief fully justifies this commemorative issue of these celebrated books.

DAVID L. GOODSTEIN
GERRY NEUGEBAUER
April 1989   California Institute of Technology

Feynman's Preface

(from *Lectures on Physics*)

These are the lectures in physics that I gave last year and the year before to the freshman and sophomore classes at Caltech. The lectures are, of course, not verbatim—they have been edited, sometimes extensively and sometimes less so. The lectures form only part of the complete course. The whole group of 180 students gathered in a big lecture room twice a week to hear these lectures and then they broke up into small groups of 15 to 20 students in recitation sections under the guidance of a teaching assistant. In addition, there was a laboratory session once a week.

The special problem we tried to get at with these lectures was to maintain the interest of the very enthusiastic and rather smart students coming out of the high schools and into Caltech. They have heard a lot about how interesting and exciting physics is—the theory of relativity, quantum mechanics, and other modern ideas. By the end of two years of our previous course, many would be very discouraged because there were really very few grand, new, modern ideas presented to them. They were made to study inclined planes, electrostatics, and so forth, and after two years it was quite stultifying. The problem was whether or not we could make a course which would save the more advanced and excited student by maintaining his enthusiasm.

The lectures here are not in any way meant to be a survey course, but are very serious. I thought to address them to the most intelligent in the class and to make sure, if possible, that even the most intelligent student was unable to completely encompass everything that was in the lectures—by putting in suggestions of applications of the ideas and concepts in various directions outside the main line of attack. For this reason, though, I tried very hard to make all the statements as accurate as possible, to point out in every case where the equations and ideas fitted into the body of physics, and how— when they learned more—things would be modified. I also felt that for such students it is important to indicate what it is that they should—if they are sufficiently clever—be able to understand by deduction from what has been said before, and what is being put in as something new. When new ideas came in, I would try either to deduce them if they were deducible, or to explain that it *was* a new idea which hadn't any basis in terms of things they had already learned and which was not supposed to be provable—but was just added in.

At the start of these lectures, I assumed that the students knew something when they came out of high school—such things as geometrical optics, simple chemistry ideas, and so on. I also didn't see that there was any reason to make the lectures in a definite order, in the sense that I would not be allowed to mention something until I was ready to discuss it in detail. There was a great deal of mention of things to come, without complete discussions. These more complete discussions would come later when the preparation became more advanced. Examples are the discussions of inductance, and of energy levels, which are at first brought in in a very qualitative way and are later developed more completely.

At the same time that I was aiming at the more active student, I also wanted to take care of the fellow for whom the extra fireworks and side applications are merely disquieting and who cannot be expected to learn most of the material in the lecture at all. For such students, I wanted there to be at least a central core or backbone of material which he *could* get. Even if he didn't understand everything in a lecture, I hoped he wouldn't get nervous. I didn't expect him to understand everything, but only the central and most direct features. It takes, of course, a certain intelligence on his part to see which are the central theorems and central ideas, and which are the

Feynman's Preface

more advanced side issues and applications which he may understand only in later years.

In giving these lectures there was one serious difficulty: in the way the course was given, there wasn't any feedback from the students to the lecturer to indicate how well the lectures were going over. This is indeed a very serious difficulty, and I don't know how good the lectures really are. The whole thing was essentially an experiment. And if I did it again I wouldn't do it the same way—I hope I *don't* have to do it again! I think, though, that things worked out—so far as the physics is concerned—quite satisfactorily in the first year.

In the second year I was not so satisfied. In the first part of the course, dealing with electricity and magnetism, I couldn't think of any really unique or different way of doing it—of any way that would be particularly more exciting than the usual way of presenting it. So I don't think I did very much in the lectures on electricity and magnetism. At the end of the second year I had originally intended to go on, after the electricity and magnetism, by giving some more lectures on the properties of materials, but mainly to take up things like fundamental modes, solutions of the diffusion equation, vibrating systems, orthogonal functions, . . . developing the first stages of what are usually called "the mathematical methods of physics." In retrospect, I think that if I were doing it again I would go back to that original idea. But since it was not planned that I would be giving these lectures again, it was suggested that it might be a good idea to try to give an introduction to the quantum mechanics—what you will find in Volume III.

It is perfectly clear that students who will major in physics can wait until their third year for quantum mechanics. On the other hand, the argument was made that many of the students in our course study physics as a background for their primary interest in other fields. And the usual way of dealing with quantum mechanics makes that subject almost unavailable for the great majority of students because they have to take so long to learn it. Yet, in its real applications—especially in its more complex applications, such as in electrical engineering and chemistry—the full machinery of the differential equation approach is not actually used. So I tried to describe the principles of quantum mechanics in a way which wouldn't require that one first know the mathematics of partial differential equations. Even for a physicist I think that is an interesting thing to try to do—to present quantum mechanics in this reverse fashion—for several reasons which may be apparent in the lectures themselves. However, I think that the experiment in the quantum mechanics part was not completely successful—in large part because I really did not have enough time at the end (I should, for instance, have had three or four more lectures in order to deal more completely with such matters as energy bands and the spatial dependence of amplitudes). Also, I had never presented the subject this way before, so the lack of feedback was particularly serious. I now believe the quantum mechanics should be given at a later time. Maybe I'll have a chance to do it again someday. Then I'll do it right.

The reason there are no lectures on how to solve problems is because there were recitation sections. Although I did put in three lectures in the first year on how to solve problems, they are not included here. Also there was a lecture on inertial guidance which certainly belongs after the lecture on rotating systems, but which was, unfortunately, omitted. The fifth and sixth lectures are actually due to Matthew Sands, as I was out of town.

The question, of course, is how well this experiment has succeeded. My own point of view—which, however, does not seem to be shared by most of the people who worked with the students—is pessimistic. I don't think I did very well by the students. When I look at the way the majority of the students handled the problems on the examinations, I think that the system is a failure. Of course, my friends point out to me that there were one or two dozen students who—very surprisingly—understood almost everything in all of the lectures, and who were quite active in working with the material and worrying about the many points in an excited and interested way. These people have now, I believe, a first-rate background in physics—and they are, after all, the ones I was trying to get at. But then, "The power of instruction is seldom of much efficacy except in those happy dispositions where it is almost superfluous." (Gibbon)

Still, I didn't want to leave any student completely behind, as perhaps I did. I think one way we could help the students more would be by putting more hard work into developing a set of problems which would elucidate some of the ideas in the lectures. Problems give a good opportunity to fill out the material of the lectures and make more realisuc, more complete, and more settled in the mind the ideas that have been exposed.

I think, however, that there isn't any solution to this problem of education other than to realize that the best teaching can be done only when there is a direct individual relationship between a student and a good teacher—a situation in which the student discusses the ideas, thinks about the things, and talks about the things. It's impossible to learn very much by simply sitting in a lecture, or even by simply doing problems that are assigned. But in our modern times we have so many students to teach that we have to try to find some substitute for the ideal. Perhaps my lectures can make some contribution. Perhaps in some small place where there are individual teachers and students, they may get some inspiration or some ideas from the lectures. Perhaps they will have fun thinking them through—or going on to develop some of the ideas further.

June 1963   RICHARD P. FEYNMAN

One

ATOMS IN MOTION

Introduction

◊ This two-year course in physics is presented from the point of view that you, the reader, are going to be a physicist. This is not necessarily the case of course, but that is what every professor in every subject assumes! If you are going to be a physicist, you will have a lot to study: two hundred years of the most rapidly developing field of knowledge that there is. So much knowledge, in fact, that you might think that you cannot learn all of it in four years, and truly you cannot; you will have to go to graduate school too!

Surprisingly enough, in spite of the tremendous amount of work that has been done for all this time it is possible to condense the enormous mass of results to a large extent—that is, to find *laws* which summarize all our knowledge. Even so, the laws are so hard to grasp that it is unfair to you to start exploring this tremendous subject without some kind of map or outline of the relationship of one part of the subject of science to another. Following these preliminary remarks, the first three chapters will therefore outline the relation of physics to the rest of the sciences, the relations of the sciences to each other, and the meaning of science, to help us develop a "feel" for the subject.

You might ask why we cannot teach physics by just giving the basic laws on page one and then showing how they work in all possible circumstances, as we do in Euclidean geometry, where we

1

state the axioms and then make all sorts of deductions. (So, not satisfied to learn physics in four years, you want to learn it in four minutes?) We cannot do it in this way for two reasons. First, we do not yet *know* all the basic laws: there is an expanding frontier of ignorance. Second, the correct statement of the laws of physics involves some very unfamiliar ideas which require advanced mathematics for their description. Therefore, one needs a considerable amount of preparatory training even to learn what the *words* mean. No, it is not possible to do it that way. We can only do it piece by piece.

Each piece, or part, of the whole of nature is always merely an *approximation* to the complete truth, or the complete truth so far as we know it. In fact, everything we know is only some kind of approximation, because *we know that we do not know all the laws* as yet. Therefore, things must be learned only to be unlearned again or, more likely, to be corrected.

The principle of science, the definition, almost, is the following: *The test of all knowledge is experiment.* Experiment is the *sole judge* of scientific "truth." But what is the source of knowledge? Where do the laws that are to be tested come from? Experiment, itself, helps to produce these laws, in the sense that it gives us hints. But also needed is *imagination* to create from these hints the great generalizations—to guess at the wonderful, simple, but very strange patterns beneath them all, and then to experiment to check again whether we have made the right guess. This imagining process is so difficult that there is a division of labor in physics: there are *theoretical* physicists who imagine, deduce, and guess at new laws, but do not experiment; and then there are *experimental* physicists who experiment, imagine, deduce, and guess.

We said that the laws of nature are approximate: that we first find the "wrong" ones, and then we find the "right" ones. Now, how can an experiment be "wrong"? First, in a trivial way: if something is wrong with the apparatus that you did not notice. But these things are easily fixed, and checked back and forth. So without snatching

Atoms in Motion

at such minor things, how *can* the results of an experiment be wrong? Only by being inaccurate. For example, the mass of an object never seems to change: a spinning top has the same weight as a still one. So a "law" was invented: mass is constant, independent of speed. That "law" is now found to be incorrect. Mass is found to increase with velocity, but appreciable increases require velocities near that of light. A *true* law is: if an object moves with a speed of less than one hundred miles a second the mass is constant to within one part in a million. In some such approximate form this is a correct law. So in practice one might think that the new law makes no significant difference. Well, yes and no. For ordinary speeds we can certainly forget it and use the simple constant-mass law as a good approximation. But for high speeds we are wrong, and the higher the speed, the more wrong we are.

Finally, and most interesting, *philosophically we are completely wrong* with the approximate law. Our entire picture of the world has to be altered even though the mass changes only by a little bit. This is a very peculiar thing about the philosophy, or the ideas, behind the laws. Even a very small effect sometimes requires profound changes in our ideas.

Now, what should we teach first? Should we teach the *correct* but unfamiliar law with its strange and difficult conceptual ideas, for example the theory of relativity, four-dimensional space-time, and so on? Or should we first teach the simple "constant-mass" law, which is only approximate, but does not involve such difficult ideas? The first is more exciting, more wonderful, and more fun, but the second is easier to get at first, and is a first step to a real understanding of the second idea. This point arises again and again in teaching physics. At different times we shall have to resolve it in different ways, but at each stage it is worth learning what is now known, how accurate it is, how it fits into everything else, and how it may be changed when we learn more.

Let us now proceed with our outline, or general map, of our understanding of science today (in particular, physics, but also of other sciences on the periphery), so that when we later concentrate on some particular point we will have some idea of the background, why that particular point is interesting, and how it fits into the big structure. So, what *is* our overall picture of the world?

Matter is made of atoms

If, in some cataclysm, all of scientific knowledge were to be destroyed, and only one sentence passed on to the next generations of creatures, what statement would contain the most information in the fewest words? I believe it is the *atomic hypothesis* (or the atomic *fact*, or whatever you wish to call it) that *all things are made of atoms—little particles that move around in perpetual motion, attracting each other when they are a little distance apart, but repelling upon being squeezed into one another*. In that one sentence, you will see, there is an *enormous* amount of information about the world, if just a little imagination and thinking are applied.

To illustrate the power of the atomic idea, suppose that we have a drop of water a quarter of an inch on the side. If we look at it very closely we see nothing but water—smooth, continuous water. Even if we magnify it with the best optical microscope available— roughly two thousand times—then the water drop will be roughly forty feet across, about as big as a large room, and if we looked rather closely, we would *still* see relatively smooth water—but here and there small football-shaped things swimming back and forth. Very interesting. These are paramecia. You may stop at this point and get so curious about the paramecia with their wiggling cilia and twisting bodies that you go no further, except perhaps to magnify the paramecia still more and see inside. This, of course, is a subject for biology, but for the present we pass on and look still more closely at the water material itself, magnifying it two thousand times again. Now the drop of water extends about fifteen miles across, and if we look very closely at it we see a kind of teeming, something which no longer has a smooth appearance—it looks something like a crowd at a football game as seen from a very great distance. In

Atoms in Motion

order to see what this teeming is about, we will magnify it another two hundred and fifty times and we will see something similar to what is shown in Fig. 1–1. This is a picture of water magnified a billion times, but idealized in several ways. In the first place, the particles are drawn in a simple manner with sharp edges, which is inaccurate. Secondly, for simplicity, they are sketched almost schematically in a two-dimensional arrangement, but of course they are moving around in three dimensions. Notice that there are two kinds of "blobs" or circles to represent the atoms of oxygen (black) and hydrogen (white), and that each oxygen has two hydrogens tied to it. (Each little group of an oxygen with its two hydrogens is called a molecule.) The picture is idealized further in that the real particles in nature are continually jiggling and bouncing, turning and twisting around one another. You will have to imagine this as a dynamic rather than a static picture. Another thing that cannot be illustrated in a drawing is the fact that the particles are "stuck together"—that they attract each other, this one pulled by that one, etc. The whole group is "glued together," so to speak. On the other hand, the particles do not squeeze through each other. If you try to squeeze two of them too close together, they repel.

The atoms are 1 or 2×10^{-8} cm in radius. Now 10^{-8} cm is called an *angstrom* (just as another name), so we say they are 1 or 2 angstroms (A) in radius. Another way to remember their size is this: if an apple is magnified to the size of the earth, then the atoms in the apple are approximately the size of the original apple.

Now imagine this great drop of water with all of these jiggling particles stuck together and tagging along with each other. The water keeps its volume; it does not fall apart, because of the attraction of the molecules for each other. If the drop is on a slope, where it can move from one place to another, the water will flow, but it does not just disappear—things do not just fly apart—because of the molecular attraction. Now the jiggling motion is what we represent as *heat*: when we increase the temperature, we increase the motion. If we heat the water, the jiggling increases and the volume between the atoms increases, and if the heating continues there

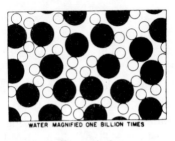

WATER MAGNIFIED ONE BILLION TIMES

Figure 1–1.

comes a time when the pull between the molecules is not enough to hold them together and they *do* fly apart and become separated from one another. Of course, this is how we manufacture steam out of water—by increasing the temperature; the particles fly apart because of the increased motion.

In Fig. 1–2 we have a picture of steam. This picture of steam fails in one respect: at ordinary atmospheric pressure there might be only a few molecules in a whole room, and there certainly would not be as many as three in this figure. Most squares this size would contain none—but we accidentally have two and a half or three in the picture (just so it would not be completely blank). Now in the case of steam we see the characteristic molecules more clearly than in the case of water. For simplicity, the molecules are drawn so that there is a 120° angle between them. In actual fact the angle is 105°3', and the distance between the center of a hydrogen and the center of the oxygen is 0.957 A, so we know this molecule very well.

Let us see what some of the properties of steam vapor or any other gas are. The molecules, being separated from one another, will bounce against the walls. Imagine a room with a number of tennis balls (a hundred or so) bouncing around in perpetual motion. When they bombard the wall, this pushes the wall away. (Of course we would have to push the wall back.) This means that the gas exerts a jittery force which our coarse senses (not being ourselves magnified a billion times) feels only as an *average push*. In order to confine a gas we must apply a pressure. Figure 1–3 shows a

Atoms in Motion

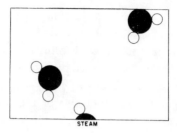

Figure 1–2.

standard vessel for holding gases (used in all textbooks), a cylinder with a piston in it. Now, it makes no difference what the shapes of water molecules are, so for simplicity we shall draw them as tennis balls or little dots. These things are in perpetual motion in all directions. So many of them are hitting the top piston all the time that to keep it from being patiently knocked out of the tank by this continuous banging, we shall have to hold the piston down by a certain force, which we call the *pressure* (really, the pressure times the area is the force). Clearly, the force is proportional to the area, for if we increase the area but keep the number of molecules per cubic centimeter the same, we increase the number of collisions with the piston in the same proportion as the area was increased.

Now let us put twice as many molecules in this tank, so as to double the density, and let them have the same speed, i.e., the same

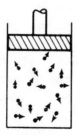

Figure 1–3.

temperature. Then, to a close approximation, the number of collisions will be doubled, and since each will be just as "energetic" as
before, the pressure is proportional to the density. If we consider
the true nature of the forces between the atoms, we would expect a
slight decrease in pressure because of the attraction between the
atoms, and a slight increase because of the finite volume they
occupy. Nevertheless, to an excellent approximation, if the density
is low enough that there are not many atoms, *the pressure is proportional to the density*.

We can also see something else: If we increase the temperature
without changing the density of the gas, i.e., if we increase the speed
of the atoms, what is going to happen to the pressure? Well, the
atoms hit harder because they are moving faster, and in addition
they hit more often, so the pressure increases. You see how simple
the ideas of atomic theory are.

Let us consider another situation. Suppose that the piston moves
inward, so that the atoms are slowly compressed into a smaller
space. What happens when an atom hits the moving piston? Evidently it picks up speed from the collision. You can try it by
bouncing a ping-pong ball from a forward-moving paddle, for
example, and you will find that it comes off with more speed than
that with which it struck. (Special example: if an atom happens to
be standing still and the piston hits it, it will certainly move.) So the
atoms are "hotter" when they come away from the piston than they
were before they struck it. Therefore all the atoms which are in the
vessel will have picked up speed. This means that *when we compress
a gas slowly, the temperature of the gas increases*. So, under slow
compression, a gas will *increase* in temperature, and under slow
expansion it will *decrease* in temperature.

We now return to our drop of water and look in another direction. Suppose that we decrease the temperature of our drop of
water. Suppose that the jiggling of the molecules of the atoms in the
water is steadily decreasing. We know that there are forces of
attraction between the atoms, so that after a while they will not be
able to jiggle so well. What will happen at very low temperatures is

Atoms in Motion

indicated in Fig. 1–4: the molecules lock into a new pattern which is *ice*. This particular schematic diagram of ice is wrong because it is in two dimensions, but it is right qualitatively. The interesting point is that the material has a *definite place for every atom*, and you can easily appreciate that if somehow or other we were to hold all the atoms at one end of the drop in a certain arrangement, each atom in a certain place, then because of the structure of interconnections, which is rigid, the other end miles away (at our magnified scale) will have a definite location. So if we hold a needle of ice at one end, the other end resists our pushing it aside, unlike the case of water, in which the structure is broken down because of the increased jiggling so that the atoms all move around in different ways. The difference between solids and liquids is, then, that in a solid the atoms are arranged in some kind of an array, called a *crystalline array*, and they do not have a random position at long distances; the position of the atoms on one side of the crystal is determined by that of other atoms millions of atoms away on the other side of the crystal. Figure 1–4 is an invented arrangement for ice, and although it contains many of the correct features of ice, it is not the true arrangement. One of the correct features is that there is a part of the symmetry that is hexagonal. You can see that if we turn the picture around an axis by 120°, the picture returns to itself. So there is a *symmetry* in the ice which accounts for the six-sided appearance of snowflakes. Another thing we can see from Fig. 1–4 is why ice

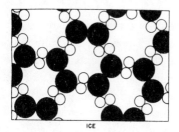

ICE

Figure 1–4.

shrinks when it melts. The particular crystal pattern of ice shown here has many "holes" in it, as does the true ice structure. When the organization breaks down, these holes can be occupied by molecules. Most simple substances, with the exception of water and type metal, *expand* upon melting, because the atoms are closely packed in the solid crystal and upon melting need more room to jiggle around, but an open structure collapses, as in the case of water.

Now although ice has a "rigid" crystalline form, its temperature can change—ice has heat. If we wish, we can change the amount of heat. What is the heat in the case of ice? The atoms are not standing still. They are jiggling and vibrating. So even though there is a definite order to the crystal—a definite structure—all of the atoms are vibrating "in place." As we increase the temperature, they vibrate with greater and greater amplitude, until they shake themselves out of place. We call this *melting*. As we decrease the temperature, the vibration decreases and decreases until, at absolute zero, there is a minimum amount of vibration that the atoms can have, but *not zero*. This minimum amount of motion that atoms can have is not enough to melt a substance, with one exception: helium. Helium merely decreases the atomic motions as much as it can, but even at absolute zero there is still enough motion to keep it from freezing. Helium, even at absolute zero, does not freeze, unless the pressure is made so great as to make the atoms squash together. If we increase the pressure, we *can* make it solidify.

Atomic processes

So much for the description of solids, liquids, and gases from the atomic point of view. However, the atomic hypothesis also describes *processes*, and so we shall now look at a number of processes from an atomic standpoint. The first process that we shall look at is associated with the surface of the water. What happens at the surface of the water? We shall now make the picture more complicated—and more realistic—by imagining that the surface is in air. Figure 1–5 shows the surface of water in air. We see the water

Atoms in Motion

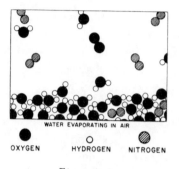

WATER EVAPORATING IN AIR

OXYGEN   HYDROGEN   NITROGEN

Figure 1–5.

molecules as before, forming a body of liquid water, but now we also see the surface of the water. Above the surface we find a number of things: First of all there are water molecules, as in steam. This is *water vapor*, which is always found above liquid water. (There is an equilibrium between the steam vapor and the water which will be described later.) In addition we find some other molecules—here two oxygen atoms stuck together by themselves, forming an *oxygen molecule*, there two nitrogen atoms also stuck together to make a nitrogen molecule. Air consists almost entirely of nitrogen, oxygen, some water vapor, and lesser amounts of carbon dioxide, argon, and other things. So above the water surface is the air, a gas, containing some water vapor. Now what is happening in this picture? The molecules in the water are always jiggling around. From time to time, one on the surface happens to be hit a little harder than usual, and gets knocked away. It is hard to see that happening in the picture because it is a *still* picture. But we can imagine that one molecule near the surface has just been hit and is flying out, or perhaps another one has been hit and is flying out. Thus, molecule by molecule, the water disappears—it evaporates. But if we *close* the vessel above, after a while we shall find a large number of molecules of water amongst the air molecules. From time to time, one of these vapor molecules comes flying down to the water and gets stuck again. So we see that what looks like a dead, uninteresting thing—a glass of water with a cover, that has been sitting there for perhaps twenty years—really contains a dynamic and interesting phenomenon which is going on all the time. To our eyes, our crude eyes, nothing is changing, but if we could see it a billion times magnified, we would see that from its own point of view it is always changing: molecules are leaving the surface, molecules are coming back.

Why do *we* see *no change*? Because just as many molecules are leaving as are coming back! In the long run "nothing happens." If we then take the top of the vessel off and blow the moist air away, replacing it with dry air, then the number of molecules leaving is just the same as it was before, because this depends on the jiggling of the water, but the number coming back is greatly reduced because there are so many fewer water molecules above the water. Therefore there are more going out than coming in, and the water evaporates. Hence, if you wish to evaporate water turn on the fan!

Here is something else: Which molecules leave? When a molecule leaves it is due to an accidental, extra accumulation of a little bit more than ordinary energy, which it needs if it is to break away from the attractions of its neighbors. Therefore, since those that leave have more energy than the average, the ones that are left have *less* average motion than they had before. So the liquid gradually *cools* if it evaporates. Of course, when a molecule of vapor comes from the air to the water below there is a sudden great attraction as the molecule approaches the surface. This speeds up the incoming molecule and results in generation of heat. So when they leave they take away heat; when they come back they generate heat. Of course when there is no net evaporation the result is nothing—the water is not changing temperature. If we blow on the water so as to maintain a continuous preponderance in the number evaporating, then the water is cooled. Hence, blow on soup to cool it!

Of course you should realize that the processes just described are more complicated than we have indicated. Not only does the water go into the air, but also, from time to time, one of the oxygen or nitrogen molecules will come in and "get lost" in the mass of water molecules, and work its way into the water. Thus the air dissolves in the water; oxygen and nitrogen molecules will work their way into

Atoms in Motion

the water and the water will contain air. If we suddenly take the air away from the vessel, then the air molecules will leave more rapidly than they come in, and in doing so will make bubbles. This is very bad for divers, as you may know.

Now we go on to another process. In Fig. 1–6 we see, from an atomic point of view, a solid dissolving in water. If we put a crystal of salt in the water, what will happen? Salt is a solid, a crystal, an organized arrangement of "salt atoms." Figure 1–7 is an illustration of the three-dimensional structure of common salt, sodium chloride. Strictly speaking, the crystal is not made of atoms, but of what we call *ions*. An ion is an atom which either has a few extra electrons or has lost a few electrons. In a salt crystal we find chlorine ions (chlorine atoms with an extra electron) and sodium ions (sodium atoms with one electron missing). The ions all stick together by electrical attraction in the solid salt, but when we put them in the water we find, because of the attractions of the negative oxygen and positive hydrogen for the ions, that some of the ions jiggle loose. In Fig. 1–6 we see a chlorine ion getting loose, and other atoms floating in the water in the form of ions. This picture was made with some care. Notice, for example, that the hydrogen ends of the water molecules are more likely to be near the chlorine ion, while near the sodium ion we are more likely to find the oxygen end, because the

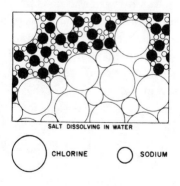

SALT DISSOLVING IN WATER

CHLORINE   SODIUM

Figure 1–6.

SIX EASY PIECES

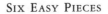

Crystal	●	○	a(Å)
Rocksalt	Na	Cl	5.64
Sylvine	K	Cl	6.28
	Ag	Cl	5.54
	Mg	O	4.20
Galena	Pb	S	5.97
	Pb	Se	6.14
	Pb	Te	6.34

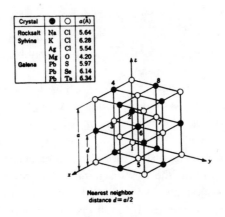

Nearest neighbor
distance $d = a/2$

Figure 1–7.

sodium is positive and the oxygen end of the water is negative, and they attract electrically. Can we tell from this picture whether the salt is *dissolving in* water or *crystallizing out* of water? Of course we *cannot* tell, because while some of the atoms are leaving the crystal other atoms are rejoining it. The process is a *dynamic* one, just as in the case of evaporation, and it depends on whether there is more or less salt in the water than the amount needed for equilibrium. By equilibrium we mean that situation in which the rate at which atoms are leaving just matches the rate at which they are coming back. If there is almost no salt in the water, more atoms leave than return, and the salt dissolves. If, on the other hand, there are too many "salt atoms," more return than leave, and the salt is crystallizing.

In passing, we mention that the concept of a *molecule* of a substance is only approximate and exists only for a certain class of substances. It is clear in the case of water that the three atoms are actually stuck together. It is not so clear in the case of sodium chloride in the solid. There is just an arrangement of sodium and chlorine ions in a cubic pattern. There is no natural way to group them as "molecules of salt."

Returning to our discussion of solution and precipitation, if we increase the temperature of the salt solution, then the rate at which

Atoms in Motion

atoms are taken away is increased, and so is the rate at which atoms are brought back. It turns out to be very difficult, in general, to predict which way it is going to go, whether more or less of the solid will dissolve. Most substances dissolve more, but some substances dissolve less, as the temperature increases.

Chemical reactions

In all of the processes which have been described so far, the atoms and the ions have not changed partners, but of course there are circumstances in which the atoms do change combinations, forming new molecules. This is illustrated in Fig. 1–8. A process in which the rearrangement of the atomic partners occurs is what we call a *chemical reaction*. The other processes so far described are called physical processes, but there is no sharp distinction between the two. (Nature does not care what we call it, she just keeps on doing it.) This figure is supposed to represent carbon burning in oxygen. In the case of oxygen, *two* oxygen atoms stick together very strongly. (Why do not *three* or even *four* stick together? That is one of the very peculiar characteristics of such atomic processes. Atoms are very special: they like certain particular partners, certain particular directions, and so on. It is the job of physics to analyze why each one wants what it wants. At any rate, two oxygen atoms form, saturated and happy, a molecule.)

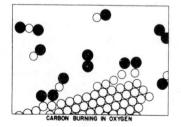

CARBON BURNING IN OXYGEN

Figure 1–8.

The carbon atoms are supposed to be in a solid crystal (which could be graphite or diamond*). Now, for example, one of the oxygen molecules can come over to the carbon, and each atom can pick up a carbon atom and go flying off in a new combination—"carbon-oxygen"—which is a molecule of the gas called carbon monoxide. It is given the chemical name CO. It is very simple: the letters "CO" are practically a picture of that molecule. But carbon attracts oxygen much more than oxygen attracts oxygen or carbon attracts carbon. Therefore in this process the oxygen may arrive with only a little energy, but the oxygen and carbon will snap together with a tremendous vengeance and commotion, and everything near them will pick up the energy. A large amount of motion energy, kinetic energy, is thus generated. This of course is *burning*; we are getting *heat* from the combination of oxygen and carbon. The heat is ordinarily in the form of the molecular motion of the hot gas, but in certain circumstances it can be so enormous that it generates *light*. That is how one gets *flames*.

In addition, the carbon monoxide is not quite satisfied. It is possible for it to attach another oxygen, so that we might have a much more complicated reaction in which the oxygen is combining with the carbon, while at the same time there happens to be a collision with a carbon monoxide molecule. One oxygen atom could attach itself to the CO and ultimately form a molecule, composed of one carbon and two oxygens, which is designated CO_2 and called carbon dioxide. If we burn the carbon with very little oxygen in a very rapid reaction (for example, in an automobile engine, where the explosion is so fast that there is not time for it to make carbon dioxide) a considerable amount of carbon monoxide is formed. In many such rearrangements, a very large amount of energy is released, forming explosions, flames, etc., depending on the reactions. Chemists have studied these arrangements of the atoms, and found that every substance is some type of *arrangement of atoms*.

---

* One *can* burn a diamond in air.

Atoms in Motion

To illustrate this idea, let us consider another example. If we go into a field of small violets, we know what "that smell" is. It is some kind of *molecule*, or arrangement of atoms, that has worked its way into our noses. First of all, *how* did it work its way in? That is rather easy. If the smell is some kind of molecule in the air, jiggling around and being knocked every which way, it might have *accidentally* worked its way into the nose. Certainly it has no particular desire to get into our nose. It is merely one helpless part of a jostling crowd of molecules, and in its aimless wanderings this particular chunk of matter happens to find itself in the nose.

Now chemists can take special molecules like the odor of violets, and analyze them and tell us the *exact arrangement* of the atoms in space. We know that the carbon dioxide molecule is straight and symmetrical: O—C—O. (That can be determined easily, too, by physical methods.) However, even for the vastly more complicated arrangements of atoms that there are in chemistry, one can, by a long, remarkable process of detective work, find the arrangements of the atoms. Figure 1–9 is a picture of the air in the neighborhood of a violet; again we find nitrogen and oxygen in the air, and water vapor. (Why is there water vapor? Because the violet is *wet*. All plants transpire.) However, we also see a "monster" composed of carbon atoms, hydrogen atoms, and oxygen atoms, which have picked a certain particular pattern in which to be arranged. It is a much more complicated arrangement than that of carbon dioxide;

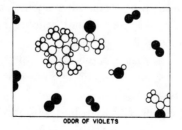

ODOR OF VIOLETS

Figure 1–9.

in fact, it is an enormously complicated arrangement. Unfortunately, we cannot picture all that is really known about it chemically, because the precise arrangement of all the atoms is actually known in three dimensions, while our picture is in only two dimensions. The six carbons which form a ring do not form a flat ring, but a kind of "puckered" ring. All of the angles and distances are known. So a chemical *formula* is merely a picture of such a molecule. When the chemist writes such a thing on the blackboard, he is trying to "draw," roughly speaking, in two dimensions. For example, we see a "ring" of six carbons, and a "chain" of carbons hanging on the end, with an oxygen second from the end, three hydrogens tied to that carbon, two carbons and three hydrogens sticking up here, etc.

How does the chemist find what the arrangement is? He mixes bottles full of stuff together, and if it turns red, it tells him that it consists of one hydrogen and two carbons tied on here; if it turns blue, on the other hand, that is not the way it is at all. This is one of the most fantastic pieces of detective work that has ever been done—organic chemistry. To discover the arrangement of the atoms in these enormously complicated arrays the chemist looks at what happens when he mixes two different substances together. The physicist could never quite believe that the chemist knew what he was talking about when he described the arrangement of the atoms. For about twenty years it has been possible, in some cases, to look at such molecules (not quite as complicated as this one, but some which contain parts of it) by a physical method, and it has been possible to locate every atom, not by looking at colors, but by *measuring where they are*. And lo and behold!, the chemists are almost always correct.

It turns out, in fact, that in the odor of violets there are three slightly different molecules, which differ only in the arrangement of the hydrogen atoms.

One problem of chemistry is to name a substance, so that we will know what it is. Find a name for this shape! Not only must the name

Atoms in Motion

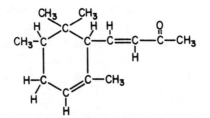

Figure 1–10. The substance pictured is α-irone.

tell the shape, but it must also tell that here is an oxygen atom, there a hydrogen—exactly what and where each atom is. So we can appreciate that the chemical names must be complex in order to be complete. You see that the name of this thing in the more complete form that will tell you the structure of it is 4-(2, 2, 3, 6 tetramethyl-5-cyclohexanyl)-3-buten-2-one, and that tells you that this is the arrangement. We can appreciate the difficulties that the chemists have, and also appreciate the reason for such long names. It is not that they wish to be obscure, but they have an extremely difficult problem in trying to describe the molecules in words!

How do we *know* that there are atoms? By one of the tricks mentioned earlier: we make the *hypothesis* that there are atoms, and one after the other results come out the way we predict, as they ought to if things *are* made of atoms. There is also somewhat more direct evidence, a good example of which is the following: The atoms are so small that you cannot see them with a light microscope—in fact, not even with an *electron* microscope. (With a light microscope you can only see things which are much bigger.) Now if the atoms are always in motion, say in water, and we put a big ball of something in the water, a ball much bigger than the atoms, the ball will jiggle around—much as in a push ball game, where a great big ball is pushed around by a lot of people. The people are pushing in various directions, and the ball moves around the field in an irregular fashion. So, in the same way, the "large ball" will move because of the inequalities of the collisions on one side to the other, from one moment to the next. Therefore, if we look at very tiny particles (colloids) in water through an excellent microscope, we see a perpetual jiggling of the particles, which is the result of the bombardment of the atoms. This is called the *Brownian motion*.

We can see further evidence for atoms in the structure of crystals. In many cases the structures deduced by x-ray analysis agree in their spatial "shapes" with the forms actually exhibited by crystals as they occur in nature. The angles between the various "faces" of a crystal agree, within seconds of arc, with angles deduced on the assumption that a crystal is made of many "layers" of atoms.

Everything is made of atoms. That is the key hypothesis. The most important hypothesis in all of biology, for example, is that *everything that animals do, atoms do.* In other words, *there is nothing that living things do that cannot be understood from the point of view that they are made of atoms acting according to the laws of physics.* This was not known from the beginning: it took some experimenting and theorizing to suggest this hypothesis, but now it is accepted, and it is the most useful theory for producing new ideas in the field of biology.

If a piece of steel or a piece of salt, consisting of atoms one next to the other, can have such interesting properties; if water—which is nothing but these little blobs, mile upon mile of the same thing over the earth—can form waves and foam, and make rushing noises and strange patterns as it runs over cement; if all of this, all the life of a stream of water, can be nothing but a pile of atoms, *how much more is possible?* If instead of arranging the atoms in some definite pattern, again and again repeated, on and on, or even forming little lumps of complexity like the odor of violets, we make an arrangement which is *always different* from place to place, with different kinds of atoms arranged in many ways, continually changing, not repeating, how much more marvelously is it possible that this thing might behave? Is it possible that that "thing" walking back and forth in front of you, talking to you, is a great glob of these atoms in

Atoms in Motion

a very complex arrangement, such that the sheer complexity of it staggers the imagination as to what it can do? When we say we are a pile of atoms, we do not mean we are *merely* a pile of atoms, because a pile of atoms which is not repeated from one to the other might well have the possibilities which you see before you in the mirror.

Two

BASIC PHYSICS

Introduction

◈ In this chapter, we shall examine the most fundamental ideas
that we have about physics—the nature of things as we see them
at the present time. We shall not discuss the history of how we know
that all these ideas are true; you will learn these details in due time.

The things with which we concern ourselves in science appear in
myriad forms, and with a multitude of attributes. For example, if we
stand on the shore and look at the sea, we see the water, the waves
breaking, the foam, the sloshing motion of the water, the sound, the
air, the winds and the clouds, the sun and the blue sky, and light;
there is sand and there are rocks of various hardness and perma-
nence, color and texture. There are animals and seaweed, hunger
and disease, and the observer on the beach; there may be even
happiness and thought. Any other spot in nature has a similar
variety of things and influences. It is always as complicated as that,
no matter where it is. Curiosity demands that we ask questions, that
we try to put things together and try to understand this multitude of
aspects as perhaps resulting from the action of a relatively small
number of elemental things and forces acting in an infinite variety of
combinations.

For example: Is the sand other than the rocks? That is, is the
sand perhaps nothing but a great number of very tiny stones?
Is the moon a great rock? If we understood rocks, would we also

23

understand the sand and the moon? Is the wind a sloshing of the air analogous to the sloshing motion of the water in the sea? What common features do different movements have? What is common to different kinds of sound? How many different colors are there? And so on. In this way we try gradually to analyze all things, to put together things which at first sight look different, with the hope that we may be able to *reduce* the number of *different* things and thereby understand them better.

A few hundred years ago, a method was devised to find partial answers to such questions. *Observation, reason,* and *experiment* make up what we call the *scientific method.* We shall have to limit ourselves to a bare description of our basic view of what is sometimes called *fundamental physics,* or fundamental ideas which have arisen from the application of the scientific method.

What do we mean by "understanding" something? We can imagine that this complicated array of moving things which constitutes "the world" is something like a great chess game being played by the gods, and we are observers of the game. We do not know what the rules of the game are; all we are allowed to do is to *watch* the playing. Of course, if we watch long enough, we may eventually catch on to a few of the rules. *The rules of the game* are what we mean by *fundamental physics.* Even if we knew every rule, however, we might not be able to understand why a particular move is made in the game, merely because it is too complicated and our minds are limited. If you play chess you must know that it is easy to learn all the rules, and yet it is often very hard to select the best move or to understand why a player moves as he does. So it is in nature, only much more so; but we may be able at least to find all the rules. Actually, we do not have all the rules now. (Every once in a while something like castling is going on that we still do not understand.) Aside from not knowing all of the rules, what we really can explain in terms of those rules is very limited, because almost all situations are so enormously complicated that we cannot follow the plays of the game using the rules, much less tell what is going to happen next. We must, therefore, limit ourselves to the more basic question

Basic Physics

of the rules of the game. If we know the rules, we consider that we "understand" the world.

How can we tell whether the rules which we "guess" at are really right if we cannot analyze the game very well? There are, roughly speaking, three ways. First, there may be situations where nature has arranged, or we arrange nature, to be simple and to have so few parts that we can predict exactly what will happen, and thus we can check how our rules work. (In one corner of the board there may be only a few chess pieces at work, and that we can figure out exactly.)

A second good way to check rules is in terms of less specific rules derived from them. For example, the rule on the move of a bishop on a chessboard is that it moves only on the diagonal. One can deduce, no matter how many moves may be made, that a certain bishop will always be on a red square. So, without being able to follow the details, we can always check our idea about the bishop's motion by finding out whether it is always on a red square. Of course it will be, for a long time, until all of a sudden we find that it is on a *black* square (what happened of course, is that in the meantime it was captured, another pawn crossed for queening, and it turned into a bishop on a black square). That is the way it is in physics. For a long time we will have a rule that works excellently in an overall way, even when we cannot follow the details, and then some time we may discover a *new rule*. From the point of view of basic physics, the most interesting phenomena are of course in the *new* places, the places where the rules do not work—not the places where they *do* work! That is the way in which we discover new rules.

The third way to tell whether our ideas are right is relatively crude but probably the most powerful of them all. That is, by rough *approximation*. While we may not be able to tell why Alekhine moves *this particular piece*, perhaps we can *roughly* understand that he is gathering his pieces around the king to protect it, more or less, since that is the sensible thing to do in the circumstances. In the same way, we can often understand nature, more or less, without being able to see what *every little piece* is doing, in terms of our understanding of the game.

At first the phenomena of nature were roughly divided into classes, like heat, electricity, mechanics, magnetism, properties of substances, chemical phenomena, light or optics, x-rays, nuclear physics, gravitation, meson phenomena, etc. However, the aim is to see *complete nature* as different aspects of *one set* of phenomena. That is the problem in basic theoretical physics, today—to *find the laws behind experiment*; to *amalgamate these classes*. Historically, we have always been able to amalgamate them, but as time goes on new things are found. We were amalgamating very well, when all of a sudden x-rays were found. Then we amalgamated some more, and mesons were found. Therefore, at any stage of the game, it always looks rather messy. A great deal is amalgamated, but there are always many wires or threads hanging out in all directions. That is the situation today, which we shall try to describe.

Some historic examples of amalgamation are the following. First, take *heat* and *mechanics*. When atoms are in motion, the more motion, the more heat the system contains, and so *heat and all temperature effects can be represented by the laws of mechanics*. Another tremendous amalgamation was the discovery of the relation between electricity, magnetism, and light, which were found to be different aspects of the same thing, which we call today the *electromagnetic field*. Another amalgamation is the unification of chemical phenomena, the various properties of various substances, and the behavior of atomic particles, which is in the *quantum mechanics of chemistry*.

The question is, of course, is it going to be possible to amalgamate *everything*, and merely discover that this world represents different aspects of *one* thing? Nobody knows. All we know is that as we go along, we find that we can amalgamate pieces, and then we find some pieces that do not fit, and we keep trying to put the jigsaw puzzle together. Whether there are a finite number of pieces, and whether there is even a border to the puzzle, is of course unknown.

Basic Physics

It will never be known until we finish the picture, if ever. What we wish to do here is to see to what extent this amalgamation process has gone on, and what the situation is at present, in understanding basic phenomena in terms of the smallest set of principles. To express it in a simple manner, *what are things made of and how few elements are there?*

Physics before 1920

It is a little difficult to begin at once with the present view, so we shall first see how things looked in about 1920 and then take a few things out of that picture. Before 1920, our world picture was something like this: The "stage" on which the universe goes is the three-dimensional *space* of geometry, as described by Euclid, and things change in a medium called *time*. The elements on the stage are *particles*, for example the atoms, which have some *properties*. First, the property of inertia: if a particle is moving it keeps on going in the same direction unless *forces* act upon it. The second element, then, is *forces*, which were then thought to be of two varieties: First, an enormously complicated, detailed kind of interaction force which held the various atoms in different combinations in a complicated way, which determined whether salt would dissolve faster or slower when we raise the temperature. The other force that was known was a long-range interaction—a smooth and quiet attraction—which varied inversely as the square of the distance, and was called *gravitation*. This law was known and was very simple. *Why* things remain in motion when they are moving, or *why* there is a law of gravitation was, of course, not known.

A description of nature is what we are concerned with here. From this point of view, then, a gas, and indeed *all* matter, is a myriad of moving particles. Thus many of the things we saw while standing at the seashore can immediately be connected. First the pressure: this comes from the collisions of the atoms with the walls or whatever; the drift of the atoms, if they are all moving in one direction on the average, is wind; the *random* internal motions are the *heat*. There are waves of excess density, where too many particles have collected, and so as they rush off they push up piles of particles farther out, and so on. This wave of excess density is *sound*. It is a tremendous achievement to be able to understand so much. Some of these things were described in the previous chapter.

What *kinds* of particles are there? There were considered to be 92 at that time: 92 different kinds of atoms were ultimately discovered. They had different names associated with their chemical properties.

The next part of the problem was, *what are the short-range forces?* Why does carbon attract one oxygen or perhaps two oxygens, but not three oxygens? What is the machinery of interaction between atoms? Is it gravitation? The answer is no. Gravity is entirely too weak. But imagine a force analogous to gravity, varying inversely with the square of the distance, but enormously more powerful and having one difference. In gravity everything attracts everything else, but now imagine that there are *two kinds* of "things," and that this new force (which is the electrical force, of course) has the property that likes *repel* but unlikes *attract*. The "thing" that carries this strong interaction is called *charge*.

Then what do we have? Suppose that we have two unlikes that attract each other, a plus and a minus, and that they stick very close together. Suppose we have another charge some distance away. Would it feel any attraction? It would feel *practically none*, because if the first two are equal in size, the attraction for the one and the repulsion for the other balance out. Therefore there is very little force at any appreciable distance. On the other hand, if we get *very close* with the extra charge, *attraction* arises, because the repulsion of likes and attraction of unlikes will tend to bring unlikes closer together and push likes farther apart. Then the repulsion will be *less* than the attraction. This is the reason why the atoms, which are constituted out of plus and minus electric charges, feel very little

Basic Physics

force when they are separated by appreciable distance (aside from gravity). When they come close together, they can "see inside" each other and rearrange their charges, with the result that they have a very strong interaction. The ultimate basis of an interaction between the atoms is *electrical*. Since this force is so enormous, all the plusses and all minuses will normally come together in as intimate a combination as they can. All things, even ourselves, are made of fine-grained, enormously strongly interacting plus and minus parts, all neatly balanced out. Once in a while, by accident, we may rub off a few minuses or a few plusses (usually it is easier to rub off minuses), and in those circumstances we find the force of electricity *unbalanced*, and we can then see the effects of these electrical attractions.

To give an idea of how much stronger electricity is than gravitation, consider two grains of sand, a millimeter across, thirty meters apart. If the force between them were not balanced, if everything attracted everything else instead of likes repelling, so that there were no cancellation, how much force would there be? There would be a force of *three million tons* between the two! You see, there is very, *very* little excess or deficit of the number of negative or positive charges necessary to produce appreciable electrical effects. This is, of course, the reason why you cannot see the difference between an electrically charged or uncharged thing—so few particles are involved that they hardly make a difference in the weight or size of an object.

With this picture the atoms were easier to understand. They were thought to have a "nucleus" at the center, which is positively electrically charged and very massive, and the nucleus is surrounded by a certain number of "electrons" which are very light and negatively charged. Now we go a little ahead in our story to remark that in the nucleus itself there were found two kinds of particles, protons and neutrons, almost of the same weight and very heavy. The protons are electrically charged and the neutrons are neutral. If we have an atom with six protons inside its nucleus, and this is surrounded by six electrons (the negative particles in the ordinary world of matter are all electrons, and these are very light compared with the protons and neutrons which make nuclei), this would be atom number six in the chemical table, and it is called carbon. Atom number eight is called oxygen, etc., because the chemical properties depend upon the electrons on the *outside*, and in fact only upon *how many* electrons there are. So the *chemical* properties of a substance depend only on a number, the number of electrons. (The whole list of elements of the chemists really could have been called 1, 2, 3, 4, 5, etc. Instead of saying "carbon," we could say "element six," meaning six electrons, but of course, when the elements were first discovered, it was not known that they could be numbered that way, and secondly, it would make everything look rather complicated. It is better to have names and symbols for these things, rather than to call everything by number.)

More was discovered about the electrical force. The natural interpretation of electrical interaction is that two objects simply attract each other: plus against minus. However, this was discovered to be an inadequate idea to represent it. A more adequate representation of the situation is to say that the existence of the positive charge, in some sense, distorts, or creates a "condition" in space, so that when we put the negative charge in, it feels a force. This potentiality for producing a force is called an *electric field*. When we put an electron in an electric field, we say it is "pulled." We then have two rules: (a) charges make a field, and (b) charges in fields have forces on them and move. The reason for this will become clear when we discuss the following phenomena: If we were to charge a body, say a comb, electrically, and then place a charged piece of paper at a distance and move the comb back and forth, the paper will respond by always pointing to the comb. If we shake it faster, it will be discovered that the paper is a little behind, *there is a delay* in the action. (At the first stage, when we move the comb rather slowly, we find a complication which is *magnetism*. Magnetic influences have to do with *charges in relative motion*, so magnetic forces and electric forces can really be attributed to one field, as two different aspects of exactly the same thing. A changing electric field cannot exist without magnetism.) If we move the charged paper farther out, the delay is greater. Then an interesting thing is observed. Although the forces between two charged objects should go inversely as the *square* of the distance, it is found, when we shake a charge, that the influence extends *very much farther out* than we would guess at first sight. That is, the effect falls off more slowly than the inverse square.

Here is an analogy: If we are in a pool of water and there is a floating cork very close by, we can move it "directly" by pushing the water with another cork. If you looked only at the two *corks*, all you would see would be that one moved immediately in response to the motion of the other—there is some kind of "*interaction*" between them. Of course, what we really do is to disturb the *water*; the *water* then disturbs the other cork. We could make up a "law" that if you pushed the water a little bit, an object close by in the water would move. If it were farther away, of course, the second cork would scarcely move, for we move the water *locally*. On the other hand, if we jiggle the cork a new phenomenon is involved, in which the motion of the water moves the water there, etc., and *waves* travel away, so that by jiggling, there is an influence *very much farther out*, an oscillatory influence, that cannot be understood from the direct interaction. Therefore the idea of direct interaction must be replaced with the existence of the water, or in the electrical case, with what we call the *electromagnetic field*.

The electromagnetic field can carry waves; some of these waves are *light*, others are used in *radio broadcasts*, but the general name is *electromagnetic waves*. These oscillatory waves can have various *frequencies*. The only thing that is really different from one wave to another is the *frequency of oscillation*. If we shake a charge back and forth more and more rapidly, and look at the effects, we get a whole series of different kinds of effects, which are all unified by specifying but one number, the number of oscillations per second. The usual "pickup" that we get from electric currents in the circuits in the walls of a building have a frequency of about one hundred

SIX EASY PIECES

Frequency in oscillations/sec	Name	Rough behavior
10^2	Electrical disturbance	Field
$5 \times 10^5 - 10^6$	Radio broadcast	
10^8	FM — TV	
10^{10}	Radar	Waves
$5 \times 10^{14} - 10^{15}$	Light	
10^{18}	X-rays	
10^{21}	γ-rays, nuclear	
10^{24}	γ-rays, "artificial"	Particle
10^{27}	γ-rays, in cosmic rays	

Table 2–1. The Electromagnetic Spectrum.

cycles per second. If we increase the frequency to 500 or 1000 kilocycles (1 kilocycle = 1000 cycles) per second, we are "on the air," for this is the frequency range which is used for radio broadcasts. (Of course it has nothing to do with the *air!* We can have radio broadcasts without any air.) If we again increase the frequency, we come into the range that is used for FM and TV. Going still further, we use certain short waves, for example for *radar.* Still higher, and we do not need an instrument to "see" the stuff, we can see it with the human eye. In the range of frequency from 5×10^{14} to 5×10^{15} cycles per second our eyes would see the oscillation of the charged comb, if we could shake it that fast, as red, blue, or violet light, depending on the frequency. Frequencies below this range are called infrared, and above it, ultraviolet. The fact that we can see in a particular frequency range makes that part of the electromagnetic spectrum no more impressive than the other parts from a physicist's standpoint, but from a human standpoint, of course, it *is* more interesting. If we go up even higher in frequency, we get x-rays. X-rays are nothing but very high-frequency light. If we go still higher, we get gamma rays. These two terms, x-rays and gamma rays, are used almost synonymously. Usually electromagnetic rays coming from nuclei are called gamma rays, while those of high energy from atoms are called x-rays, but at the same frequency they are indistinguishable physically, no matter what their source. If we go to still higher frequencies, say to 10^{24} cycles per second, we find that we can make those waves artificially, for example with the synchrotron here at Caltech. We can find electromagnetic waves with stupendously high frequencies—with even a thousand times more rapid oscillation—in the waves found in *cosmic rays*. These waves cannot be controlled by us.

Quantum physics

Having described the idea of the electromagnetic field, and that this field can carry waves, we soon learn that these waves actually behave in a strange way which seems very unwavelike. At higher frequencies they behave much more like *particles!* It is *quantum mechanics*, discovered just after 1920, which explains this strange behavior. In the years before 1920, the picture of space as a three-dimensional space, and of time as a separate thing, was changed by Einstein, first into a combination which we call space-time, and then still further into a *curved* space-time to represent gravitation. So the "stage" is changed into space-time, and gravitation is presumably a modification of space-time. Then it was also found that the rules for the motions of particles were incorrect. The mechanical rules of "inertia" and "forces" are *wrong*—Newton's laws are *wrong*—in the world of atoms. Instead, it was discovered that things on a small scale behave *nothing like* things on a large scale. That is what makes physics difficult—and very interesting. It is hard because the way things behave on a small scale is so "unnatural"; we have no direct experience with it. Here things behave like nothing we know of, so that it is impossible to describe this behavior in any other than analytic ways. It is difficult, and takes a lot of imagination.

Quantum mechanics has many aspects. In the first place, the idea that a particle has a definite location and a definite speed is no longer allowed; that is wrong. To give an example of how wrong classical physics is, there is a rule in quantum mechanics that says that one cannot know both where something is and how fast it is moving. The uncertainty of the momentum and the uncertainty of the position are complementary, and the product of the two is constant. We can write the law like this: $\Delta x \, \Delta p \geq h/2\pi$, but we shall explain it in more detail later. This rule is the explanation of a very mysterious paradox: if the atoms are made out of plus and minus charges, why don't the minus charges simply sit on top of the plus charges (they attract each other) and get so close as to completely cancel them out? *Why are atoms so big?* Why is the nucleus at the center with the electrons around it? It was first thought that this was because the nucleus was so big; but no, the nucleus is *very small*. An atom has a diameter of about 10^{-8} cm. The nucleus has a diameter of about 10^{-13} cm. If we had an atom and wished to see the nucleus, we would have to magnify it until the whole atom was the size of a large room, and then the nucleus would be a bare speck which you could just about make out with the eye, but very nearly *all the weight* of the atom is in that infinitesimal *nucleus*. What keeps the electrons from simply falling in? This principle: If they were in the nucleus, we would know their position precisely, and the uncertainty principle would then require that they have a very *large* (but uncertain) momentum, i.e., a very large *kinetic energy*. With this energy they would break away from the nucleus. They make a compromise: they leave themselves a little room for this uncertainty and then jiggle with a certain amount of minimum motion in accordance with this rule. (Remember that when a crystal is cooled to absolute zero, we said that the atoms do not stop moving, they still jiggle. Why? If they stopped moving, we would know where they were and that they had zero motion, and that is against the uncertainty principle. We cannot know where they are and how fast they are moving, so they must be continually wiggling in there!)

Another most interesting change in the ideas and philosophy of

Basic Physics

science brought about by quantum mechanics is this: it is not possible to predict *exactly* what will happen in any circumstance. For example, it is possible to arrange an atom which is ready to emit light, and we can measure when it has emitted light by picking up a photon particle, which we shall describe shortly. We cannot, however, predict *when* it is going to emit the light or, with several atoms, *which one* is going to. You may say that this is because there are some internal "wheels" which we have not looked at closely enough. No, there *are* no internal wheels; nature, as we understand it today, behaves in such a way that it is *fundamentally impossible* to make a precise prediction of *exactly what will happen* in a given experiment. This is a horrible thing; in fact, philosophers have said before that one of the fundamental requisites of science is that whenever you set up the same conditions, the same thing must happen. This is simply *not true*, it is *not* a fundamental condition of science. The fact is that the same thing does not happen, that we can find only an average, statistically, as to what happens. Nevertheless, science has not completely collapsed. Philosophers, incidentally, say a great deal about what is *absolutely necessary* for science, and it is always, so far as one can see, rather naive, and probably wrong. For example, some philosopher or other said it is fundamental to the scientific effort that if an experiment is performed in, say, Stockholm, and then the same experiment is done in, say, Quito, the *same results* must occur. That is quite false. It is not necessary that *science* do that; it may be a *fact of experience*, but it is not necessary. For example, if one of the experiments is to look out at the sky and see the aurora borealis in Stockholm, you do not see it in Quito; that is a different phenomenon. "But," you say, "that is something that has to do with the outside; can you close yourself up in a box in Stockholm and pull down the shade and get any difference?" Surely. If we take a pendulum on a universal joint, and pull it out and let go, then the pendulum will swing almost in a plane, but not quite. Slowly the plane keeps changing in Stockholm, but not in Quito.

Basic Physics

rule which assumes that the particle-wave idea of quantum mechanics is valid.

Thus we have a new view of electromagnetic interaction. We have a new kind of *particle* to add to the electron, the proton, and the neutron. That new particle is called a *photon*. The new view of the interaction of electrons and protons that is electromagnetic theory, but with everything quantum-mechanically correct, is called *quantum electrodynamics*. This fundamental theory of the interaction of light and matter, or electric field and charges, is our greatest success so far in physics. In this one theory we have the basic rules for all ordinary phenomena except for gravitation and nuclear processes. For example, out of quantum electrodynamics come all known electrical, mechanical, and chemical laws: the laws for the collision of billiard balls, the motions of wires in magnetic fields, the specific heat of carbon monoxide, the color of neon signs, the density of salt, and the reactions of hydrogen and oxygen to make water are all consequences of this one law. All these details can be worked out if the situation is simple enough for us to make an approximation, which is almost never, but often we can understand more or less what is happening. At the present time no exceptions are found to the quantum-electrodynamic laws outside the nucleus, and there we do not know whether there is an exception because we simply do not know what is going on in the nucleus.

In principle, then, quantum electrodynamics is the theory of all chemistry, and of life, if life is ultimately reduced to chemistry and therefore just to physics because chemistry is already reduced (the part of physics which is involved in chemistry being already known). Furthermore, the same quantum electrodynamics, this great thing, predicts a lot of new things. In the first place, it tells the properties of very high-energy photons, gamma rays, etc. It predicted another very remarkable thing: besides the electron, there should be another particle of the same mass, but of opposite charge, called a *positron*, and these two, coming together, could annihilate each other with the emission of light or gamma rays. (After all, light and gamma rays are all the same, they are just different points on a frequency scale.) The generalization of this, that for each particle there is an antiparticle, turns out to be true. In the case of electrons, the antiparticle has another name—it is called a positron, but for most other particles, it is called anti-so-and-so, like antiproton or antineutron. In quantum electrodynamics, *two numbers* are put in and most of the other numbers in the world are supposed to come out. The two numbers that are put in are called the mass of the electron and the charge of the electron. Actually, that is not quite true, for we have a whole set of numbers for chemistry which tells how heavy the nuclei are. That leads us to the next part.

Nuclei and particles

What are the nuclei made of, and how are they held together? It is found that the nuclei are held together by enormous forces. When these are released, the energy released is tremendous compared with chemical energy, in the same ratio as the atomic bomb explosion is to a TNT explosion, because, of course, the atomic bomb has to do with changes inside the nucleus, while the explosion of TNT has to do with the changes of the electrons on the outside of the atoms. The question is, what are the forces which hold the protons and neutrons together in the nucleus? Just as the electrical interaction can be connected to a particle, a photon, Yukawa suggested that the forces between neutrons and protons also have a field of some kind, and that when this field jiggles it behaves like a particle. Thus there could be some other particles in the world besides protons and neutrons, and he was able to deduce the properties of these particles from the already known characteristics of nuclear forces. For example, he predicted they should have a mass of two or three hundred times that of an electron; and lo and behold, in cosmic rays there was discovered a particle of the right mass! But it later turned out to be the wrong particle. It was called a μ-meson, or muon.

Basic Physics

However, a little while later, in 1947 or 1948, another particle was found, the π-meson, or pion, which satisfied Yukawa's criterion. Besides the proton and the neutron, then, in order to get nuclear forces we must add the pion. Now, you say, "Oh great!, with this theory we make quantum nucleodynamics using the pions just like Yukawa wanted to do, and see if it works, and everything will be explained." Bad luck. It turns out that the calculations that are involved in this theory are so difficult that no one has ever been able to figure out what the consequences of the theory are, or to check it against experiment, and this has been going on now for almost twenty years!

So we are stuck with a theory, and we do not know whether it is right or wrong, but we do know that it is a *little* wrong, or at least incomplete. While we have been dawdling around theoretically, trying to calculate the consequences of this theory, the experimentalists have been discovering some things. For example, they had already discovered this μ-meson or muon, and we do not yet know where it fits. Also, in cosmic rays, a large number of other "extra" particles were found. It turns out that today we have approximately thirty particles, and it is very difficult to understand the relationships of all these particles, and what nature wants them for, or what the connections are from one to another. We do not today understand these various particles as different aspects of the same thing, and the fact that we have so many unconnected particles is a representation of the fact that we have so much unconnected information without a good theory. After the great successes of quantum electrodynamics, there is a certain amount of knowledge of nuclear physics which is rough knowledge, sort of half experience and half theory, assuming a type of force between protons and neutrons and seeing what will happen, but not really understanding where the force comes from. Aside from that, we have made very little progress. We have collected an enormous number of chemical elements. In the chemical case, there suddenly appeared a relationship among these elements which was unexpected, and which is embodied in the periodic table of Mendeléev. For example, sodium and potassium are about the same in their chemical properties and are found in the same column in the Mendeléev chart. We have been seeking a Mendeléev-type chart for the new particles. One such chart of the new particles was made independently by Gell-Mann in the U.S.A. and Nishijima in Japan. The basis of their classification is a new number, like the electric charge, which can be assigned to each particle, called its "strangeness," S. This number is conserved, like the electric charge, in reactions which take place by nuclear forces.

In Table 2–2 are listed all the particles. We cannot discuss them much at this stage, but the table will at least show you how much we do not know. Underneath each particle its mass is given in a certain unit, called the Mev. One Mev is equal to 1.782×10^{-27} gram. The reason this unit was chosen is historical, and we shall not go into it now. More massive particles are put higher up on the chart; we see that a neutron and a proton have almost the same mass. In vertical columns we have put the particles with the same electrical charge, all neutral objects in one column, all positively charged ones to the right of this one, and all negatively charged objects to the left.

Particles are shown with a solid line and "resonances" with a dashed one. Several particles have been omitted from the table. These include the important zero-mass, zero-charge particles, the photon and the graviton, which do not fall into the baryon-meson-lepton classification scheme, and also some of the newer resonances (K^*, φ, η). The antiparticles of the mesons are listed in the table, but the antiparticles of the leptons and baryons would have to be listed in another table which would look exactly like this one reflected on the zero-charge column. Although all of the particles except the electron, neutrino, photon, graviton, and proton are unstable, decay products have been shown only for the resonances. Strangeness assignments are not applicable for leptons, since they do not interact strongly with nuclei.

All particles which are together with the neutrons and protons are called *baryons*, and the following ones exist: There is a "lambda," with a mass of 1154 Mev, and three others, called sigmas, minus, neutral, and plus, with several masses almost the same. There are groups or multiplets with almost the same mass, within one or two percent. Each particle in a multiplet has the same strangeness. The first multiplet is the proton-neutron doublet, and then there is a singlet (the lambda), then the sigma triplet, and finally the xi doublet. Very recently, in 1961, even a few more particles were found. Or *are* they particles? They live so short a time, they disintegrate almost instantaneously, as soon as they are formed, that we do not know whether they should be considered as new particles, or some kind of "resonance" interaction of a certain definite energy between the Λ and π products into which they disintegrate.

In addition to the baryons the other particles which are involved in the nuclear interaction are called *mesons*. There are first the pions, which come in three varieties, positive, negative, and neutral; they form another multiplet. We have also found some new things called K-mesons, and they occur as a doublet, K^+ and K°. Also, every particle has its antiparticle, unless a particle *is its own* antiparticle. For example, the π^- and the π^+ are antiparticles, but the π° is its own antiparticle. The K^- and K^+ are antiparticles, and the K° and $\overline{K}^\circ$. In addition, in 1961 we also found some more mesons or *maybe* mesons which disintegrate almost immediately. A thing called ω which goes into three pions has a mass 780 on this scale, and somewhat less certain is an object which disintegrates into two pions. These particles, called mesons and baryons, and the antiparticles of the mesons are on the same chart, but the antiparticles of the baryons must be put on another chart, "reflected" through the charge-zero column.

Just as Mendeléev's chart was very good, except for the fact that there were a number of rare earth elements which were hanging out loose from it, so we have a number of things hanging out loose from this chart—particles which do not interact strongly in nuclei, have nothing to do with a nuclear interaction, and do not have a strong interaction (I mean the powerful kind of interaction of nuclear energy). These are called leptons, and they are the following: there

Basic Physics

is the electron, which has a very small mass on this scale, only 0.510 Mev. Then there is that other, the μ-meson, the muon, which has a mass much higher, 206 times as heavy as an electron. So far as we can tell, by all experiments so far, the difference between the electron and the muon is nothing but the mass. Everything works exactly the same for the muon as for the electron, except that one is heavier than the other. Why is there another one heavier; what is the use for it? We do not know. In addition, there is a lepton which is neutral, called a neutrino, and this particle has zero mass. In fact, it is now known that there are *two* different kinds of neutrinos, one related to electrons and the other related to muons.

Finally, we have two other particles which do not interact strongly with the nuclear ones: one is a photon, and perhaps, if the field of gravity also has a quantum-mechanical analog (a quantum theory of gravitation has not yet been worked out), then there will be a particle, a graviton, which will have zero mass.

What is this "zero mass"? The masses given here are the masses of the particles *at rest*. The fact that a particle has zero mass means, in a way, that it cannot *be* at *rest*. A photon is never at rest, it is always moving at 186,000 miles a second. We will understand more what mass means when we understand the theory of relativity, which will come in due time.

Thus we are confronted with a large number of particles, which together seem to be the fundamental constituents of matter. Fortunately, these particles are not *all* different in their *interactions* with one another. In fact, there seem to be just *four kinds* of interaction between particles which, in the order of decreasing strength, are the nuclear force, electrical interactions, the beta-decay interaction, and gravity. The photon is coupled to all charged particles and the strength of the interaction is measured by some number, which is $\frac{1}{137}$. The detailed law of this coupling is known, that is quantum electrodynamics. Gravity is coupled to all *energy*, but its coupling is extremely weak, much weaker than that of electricity. This law is also known. Then there are the so-called weak decays—beta decay, which causes the neutron to disintegrate into proton, electron, and neutrino, relatively slowly. This law is only partly known. The so-called strong interaction, the meson-baryon interaction, has a strength of 1 in this scale, and the law is completely unknown, although there are a number of known rules, such as that the number of baryons does not change in any reaction.

Coupling	Strength*	Law
Photon to charged particles	$\sim 10^{-2}$	Law known
Gravity to all energy	$\sim 10^{-40}$	Law known
Weak decays	$\sim 10^{-5}$	Law partly known
Mesons to baryons	~ 1	Law unknown (some rules known)

Table 2–3. Elementary Interactions.

This then, is the horrible condition of our physics today. To summarize it, I would say this: outside the nucleus, we seem to know all; inside it, quantum mechanics is valid—the principles of quantum mechanics have not been found to fail. The stage on which we put all of our knowledge, we would say, is relativistic space-time; perhaps gravity is involved in space-time. We do not know how the universe got started, and we have never made experiments which check our ideas of space and time accurately, below some tiny distance, so we only *know* that our ideas work above that distance. We should also add that the rules of the game are the quantum-mechanical principles, and those principles apply, so far as we can tell, to the new particles as well as to the old. The origin of the forces in nuclei leads us to new particles, but unfortunately they appear in

* The "strength" is a dimensionless measure of the coupling constant involved in each interaction ($\sim$ means "approximately").

great profusion and we lack a complete understanding of their interrelationship, although we already know that there are some very surprising relationships among them. We seem gradually to be groping toward an understanding of the world of sub-atomic particles, but we really do not know how far we have yet to go in this task.

Three

THE RELATION OF PHYSICS TO OTHER SCIENCES

Introduction

▷ Physics is the most fundamental and all-inclusive of the sciences, and has had a profound effect on all scientific development. In fact, physics is the present-day equivalent of what used to be called *natural philosophy*, from which most of our modern sciences arose. Students of many fields find themselves studying physics because of the basic role it plays in all phenomena. In this chapter we shall try to explain what the fundamental problems in the other sciences are, but of course it is impossible in so small a space really to deal with the complex, subtle, beautiful matters in these other fields. Lack of space also prevents our discussing the relation of physics to engineering, industry, society, and war, or even the most remarkable relationship between mathematics and physics. (Mathematics is not a science from our point of view, in the sense that it is not a *natural* science. The test of its validity is not experiment.) We must, incidentally, make it clear from the beginning that if a thing is not a science, it is not necessarily bad. For example, love is not a science. So, if something is said not to be a science, it does not mean that there is something wrong with it; it just means that it is not a science.

Chemistry

The science which is perhaps the most deeply affected by physics is chemistry. Historically, the early days of chemistry dealt almost entirely with what we now call inorganic chemistry, the chemistry of substances which are not associated with living things. Considerable analysis was required to discover the existence of the many elements and their relationships—how they make the various relatively simple compounds found in rocks, earth, etc. This early chemistry was very important for physics. The interaction between the two sciences was very great because the theory of atoms was substantiated to a large extent by experiments in chemistry. The theory of chemistry, i.e., of the reactions themselves, was summarized to a large extent in the periodic chart of Mendeléev, which brings out many strange relationships among the various elements, and it was the collection of rules as to which substance is combined with which, and how, that constituted inorganic chemistry. All these rules were ultimately explained in principle by quantum mechanics, so that theoretical chemistry is in fact physics. On the other hand, it must be emphasized that this explanation is *in principle*. We have already discussed the difference between knowing the rules of the game of chess, and being able to play. So it is that we may know the rules, but we cannot play very well. It turns out to be very difficult to predict precisely what will happen in a given chemical reaction; nevertheless, the deepest part of theoretical chemistry must end up in quantum mechanics.

There is also a branch of physics and chemistry which was developed by both sciences together, and which is extremely important. This is the method of statistics applied in a situation in which there are mechanical laws, which is aptly called *statistical mechanics*. In any chemical situation a large number of atoms are involved, and we have seen that the atoms are all jiggling around in a very random and complicated way. If we could analyze each collision, and be able to follow in detail the motion of each molecule, we might hope to figure out what would happen, but the many numbers needed to keep track of all these molecules exceeds so enormously the capacity of any computer, and certainly the capacity of the mind, that it was important to develop a method for dealing with such complicated situations. Statistical mechanics, then, is the science of the phenomena of heat, or thermodynamics. Inorganic chemistry is, as a science, now reduced essentially to what are called physical chemistry and quantum chemistry; physical chemistry to study the rates at which reactions occur and what is happening in detail (How do the molecules hit? Which pieces fly off first?, etc.), and quantum chemistry to help us understand what happens in terms of the physical laws.

The other branch of chemistry is *organic chemistry*, the chemistry of the substances which are associated with living things. For a time it was believed that the substances which are associated with living things were so marvelous that they could not be made by hand, from inorganic materials. This is not at all true—they are just the same as the substances made in inorganic chemistry, but more complicated arrangements of atoms are involved. Organic chemistry obviously has a very close relationship to the biology which supplies its substances, and to industry, and furthermore, much physical chemistry and quantum mechanics can be applied to organic as well as to inorganic compounds. However, the main problems of organic chemistry are not in these aspects, but rather in the analysis and synthesis of the substances which are formed in biological systems, in living things. This leads imperceptibly, in steps, toward biochemistry, and then into biology itself, or molecular biology.

Biology

Thus we come to the science of *biology*, which is the study of living things. In the early days of biology, the biologists had to deal with the purely descriptive problem of finding out *what* living things there were, and so they just had to count such things as the hairs of the limbs of fleas. After these matters were worked out with a great deal of interest, the biologists went into the *machinery* inside the living bodies, first from a gross standpoint, naturally, because it takes some effort to get into the finer details.

There was an interesting early relationship between physics and biology in which biology helped physics in the discovery of the *conservation of energy*, which was first demonstrated by Mayer in connection with the amount of heat taken in and given out by a living creature.

If we look at the processes of biology of living animals more closely, we see *many* physical phenomena: the circulation of blood, pumps, pressure, etc. There are nerves: we know what is happening when we step on a sharp stone, and that somehow or other the information goes from the leg up. It is interesting how that happens. In their study of nerves, the biologists have come to the conclusion that nerves are very fine tubes with a complex wall which is very thin; through this wall the cell pumps ions, so that there are positive ions on the outside and negative ions on the inside, like a capacitor. Now this membrane has an interesting property; if it "discharges" in one place, i.e., if some of the ions were able to move through one place, so that the electric voltage is reduced there, that electrical influence makes itself felt on the ions in the neighborhood, and it affects the membrane in such a way that it lets the ions through at neighboring points also. This in turn affects it farther along, etc., and so there is a wave of "penetrability" of the membrane which runs down the fiber when it is "excited" at one end by stepping on the sharp stone. This wave is somewhat analogous to a long sequence of vertical dominoes; if the end one is pushed over, that one pushes the next, etc. Of course this will transmit only one message unless the dominoes are set up again; and similarly in the nerve cell, there are processes which pump the ions slowly out again, to get the nerve ready for the next impulse. So it is that we know what we are doing (or at least where we are). Of course the electrical effects associated with

The Relation of Physics to Other Sciences

this nerve impulse can be picked up with electrical instruments, and because there *are* electrical effects, obviously the physics of electrical effects has had a great deal of influence on understanding the phenomenon.

The opposite effect is that, from somewhere in the brain, a message is sent out along a nerve. What happens at the end of the nerve? There the nerve branches out into fine little things, connected to a structure near a muscle, called an endplate. For reasons which are not exactly understood, when the impulse reaches the end of the nerve, little packets of a chemical called acetylcholine are shot off (five or ten molecules at a time) and they affect the muscle fiber and make it contract—how simple! What makes a muscle contract? A muscle is a very large number of fibers close together, containing two different substances, myosin and actomyosin, but the machinery by which the chemical reaction induced by acetylcholine can modify the dimensions of the molecule is not yet known. Thus the fundamental processes in the muscle that make mechanical motions are not known.

Biology is such an enormously wide field that there are hosts of other problems that we cannot mention at all—problems on how vision works (what the light does in the eye), how hearing works, etc. (The way in which *thinking* works we shall discuss later under psychology.) Now, these things concerning biology which we have just discussed are, from a biological standpoint, really not fundamental, at the bottom of life, in the sense that even if we understood them we still would not understand life itself. To illustrate: the men who study nerves feel their work is very important, because after all you cannot have animals without nerves. But you *can* have *life* without nerves. Plants have neither nerves nor muscles, but they are working, they are alive, just the same. So for the fundamental problems of biology we must look deeper; when we do, we discover that all living things have a great many characteristics in common. The most common feature is that they are made of *cells*, within each of which is complex machinery for doing things chemically. In plant cells, for example, there is machinery for picking up light and generating sucrose, which is consumed in the dark to keep the plant alive. When the plant is eaten the sucrose itself generates in the animal a series of chemical reactions very closely related to photosynthesis (and its opposite effect in the dark) in plants.

In the cells of living systems there are many elaborate chemical reactions, in which one compound is changed into another and another. To give some impression of the enormous efforts that have gone into the study of biochemistry, the chart in Fig. 3–1 summarizes our knowledge to date on just one small part of the many series of reactions which occur in cells, perhaps a percent or so of it.

Here we see a whole series of molecules which change from one to another in a sequence or cycle of rather small steps. It is called the Krebs cycle, the respiratory cycle. Each of the chemicals and each of the steps is fairly simple, in terms of what change is made in the molecule, but—and this is a centrally important discovery in biochemistry—these changes are *relatively difficult to accomplish in a laboratory*. If we have one substance and another very similar substance, the one does not just turn into the other, because the two forms are usually separated by an energy barrier or "hill." Consider this analogy: If we wanted to take an object from one place to another, at the same level but on the other side of a hill, we could push it over the top, but to do so requires the addition of some energy. Thus most chemical reactions do not occur, because there is what is called an *activation energy* in the way. In order to add an extra atom to our chemical requires that we get it *close* enough that some rearrangement can occur; then it will stick. But if we cannot give it enough energy to get it close enough, it will not go to completion, it will just go part way up the "hill" and back down again. However, if we could literally take the molecules in our hands and push and pull the atoms around in such a way as to open a hole to let the new atom in, and then let it snap back, we would have

The Relation of Physics to Other Sciences

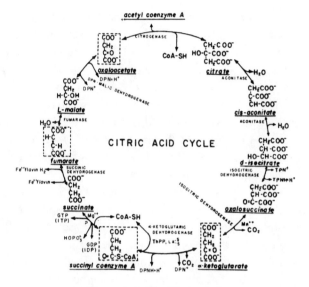

Figure 3–1. The Krebs cycle.

found another way, *around* the hill, which would not require extra energy, and the reaction would go easily. Now there actually *are*, in the cells, *very* large molecules, much larger than the ones whose changes we have been describing, which in some complicated way hold the smaller molecules just right, so that the reaction can occur easily. These very large and complicated things are called *enzymes*. (They were first called ferments, because they were originally discovered in the fermentation of sugar. In fact, some of the first reactions in the cycle were discovered there.) In the presence of an enzyme the reaction will go.

An enzyme is made of another substance called *protein*. Enzymes are very big and complicated, and each one is different, each being built to control a certain special reaction. The names of the enzymes are written in Fig. 3–1 at each reaction. (Sometimes the same enzyme may control two reactions.) We emphasize that the enzymes themselves are not involved in the reaction directly. They do not change; they merely let an atom go from one place to another. Having done so, the enzyme is ready to do it to the next molecule, like a machine in a factory. Of course, there must be a supply of certain atoms and a way of disposing of other atoms. Take hydrogen, for example: there are enzymes which have special units on them which carry the hydrogen for all chemical reactions. For example, there are three or four hydrogen-reducing enzymes which are used all over our cycle in different places. It is interesting that the machinery which liberates some hydrogen at one place will take that hydrogen and use it somewhere else.

The most important feature of the cycle of Fig. 3–1 is the transformation from GDP to GTP (guanadine-di-phosphate to guanadine-tri-phosphate) because the one substance has much more energy in it than the other. Just as there is a "box" in certain enzymes for carrying hydrogen atoms around, there are special *energy*-carrying "boxes" which involve the triphosphate group. So, GTP has more energy than GDP and if the cycle is going one way, we are producing molecules which have extra energy and which can go drive some other cycle which *requires* energy, for example the contraction of muscle. The muscle will not contract unless there is GTP. We can take muscle fiber, put it in water, and add GTP, and the fibers contract, changing GTP to GDP if the right enzymes are present. So the real system is in the GDP-GTP transformation; in the dark the GTP which has been stored up during the day is used to run the whole cycle around the other way. An enzyme, you see, does not care in which direction the reaction goes, for if it did it would violate one of the laws of physics.

Physics is of great importance in biology and other sciences for still another reason, that has to do with *experimental techniques*. In fact, if it were not for the great development of experimental physics, these biochemistry charts would not be known today. The reason is that the most useful tool of all for analyzing this fantastically complex system is to *label* the atoms which are used in the reactions. Thus, if we could introduce into the cycle some carbon

The Relation of Physics to Other Sciences

dioxide which has a "green mark" on it, and then measure after three seconds where the green mark is, and again measure after ten seconds, etc., we could trace out the course of the reactions. What are the "green marks"? They are different *isotopes*. We recall that the chemical properties of atoms are determined by the number of *electrons*, not by the mass of the nucleus. But there can be, for example in carbon, six neutrons or seven neutrons, together with the six protons which all carbon nuclei have. Chemically, the two atoms C^{12} and C^{13} are the same, but they differ in weight and they have different nuclear properties, and so they are distinguishable. By using these isotopes of different weights, or even radioactive isotopes like C^{14}, which provide a more sensitive means for tracing very small quantities, it is possible to trace the reactions.

Now, we return to the description of enzymes and proteins. All proteins are not enzymes, but all enzymes are proteins. There are many proteins, such as the proteins in muscle, the structural proteins which are, for example, in cartilage and hair, skin, etc., that are not themselves enzymes. However, proteins are a very characteristic substance of life: first of all they make up all the enzymes, and second, they make up much of the rest of living material. Proteins have a very interesting and simple structure. They are a series, or chain, of different *amino acids*. There are twenty different amino acids, and they all can combine with each other to form chains in which the backbone is CO-NH, etc. Proteins are nothing but chains of various ones of these twenty amino acids. Each of the amino acids probably serves some special purpose. Some, for example, have a sulphur atom at a certain place; when two sulphur atoms are in the same protein, they form a bond, that is, they tie the chain together at two points and form a loop. Another has extra oxygen atoms which make it an acidic substance, another has a basic characteristic. Some of them have big groups hanging out to one side, so that they take up a lot of space. One of the amino acids, called prolene, is not really an amino acid, but imino acid. There is a slight difference, with the result that when prolene is in the chain, there is a kink in the chain. If we wished to manufacture a particular protein, we would give these instructions: put one of those sulphur hooks here; next, add something to take up space; then attach something to put a kink in the chain. In this way, we will get a complicated-looking chain, hooked together and having some complex structure; this is presumably just the manner in which all the various enzymes are made. One of the great triumphs in recent times (since 1960) was at last to discover the exact spatial atomic arrangement of certain proteins, which involve some fifty-six or sixty amino acids in a row. Over a thousand atoms (more nearly two thousand, if we count the hydrogen atoms) have been located in a complex pattern in two proteins. The first was hemoglobin. One of the sad aspects of this discovery is that we cannot see anything from the pattern; we do not understand why it works the way it does. Of course, that is the next problem to be attacked.

Another problem is how do the enzymes know what to be? A red-eyed fly makes a red-eyed fly baby, and so the information for the whole pattern of enzymes to make red pigment must be passed from one fly to the next. This is done by a substance in the nucleus of the cell, not a protein, called DNA (short for desoxyribose nucleic acid). This is the key substance which is passed from one cell to another (for instance, sperm cells consist mostly of DNA) and carries the information as to how to make the enzymes. DNA is the "blueprint." What does the blueprint look like and how does it work? First, the blueprint must be able to reproduce itself. Secondly, it must be able to instruct the protein. Concerning the reproduction, we might think that this proceeds like cell reproduction. Cells simply grow bigger and then divide in half. Must it be thus with DNA molecules, then, that they too grow bigger and divide in half? Every *atom* certainly does not grow bigger and divide in half! No, it is impossible to reproduce a molecule except by some more clever way.

The structure of the substance DNA was studied for a long time, first chemically to find the composition, and then with x-rays to find the pattern in space. The result was the following remarkable discovery: The DNA molecule is a pair of chains, twisted upon each other. The backbone of each of these chains, which are analogous to the chains of proteins but chemically quite different, is a series of sugar and phosphate groups, as shown in Fig. 3–2. Now we see how the chain can contain instructions, for if we could split this chain down the middle, we would have a series *BAADC* . . . and every living thing could have a different series. Thus perhaps, in some way, the specific *instructions* for the manufacture of proteins are contained in the specific *series* of the DNA.

Attached to each sugar along the line, and linking the two chains together, are certain pairs of cross-links. However, they are not all of the same kind; there are four kinds, called adenine, thymine, cytosine, and guanine, but let us call them *A, B, C,* and *D.* The interesting thing is that only certain pairs can sit opposite each other, for example *A* with *B* and *C* with *D.* These pairs are put on the two chains in such a way that they "fit together," and have a strong energy of interaction. However, *C* will not fit with *A,* and *B* will not fit with *C;* they will only fit in pairs, *A* against *B* and *C* against *D.* Therefore if one is *C,* the other must be *D,* etc. Whatever the letters may be in one chain, each one must have its specific complementary letter on the other chain.

What then about reproduction? Suppose we split this chain in two. How can we make another one just like it? If, in the substances of the cells, there is a manufacturing department which brings up phosphate, sugar, and *A, B, C, D* units not connected in a chain, the only ones which will attach to our split chain will be the correct ones, the complements of *BAADC* . . . , namely, *ABBCD* . . . Thus what happens is that the chain splits down the middle during cell division, one half ultimately to go with one cell, the other half to end up in the other cell; when separated, a new complementary chain is made by each half-chain.

Next comes the question, precisely how does the order of the *A, B, C, D* units determine the arrangement of the amino acids in the protein? This is the central unsolved problem in biology today. The

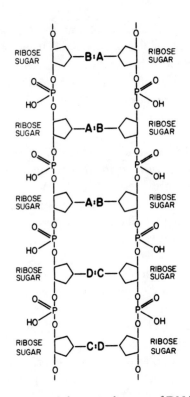

Figure 3–2. Schematic diagram of DNA.

first clues, or pieces of information, however, are these: There are in the cell tiny particles called microsomes, and it is now known that that is the place where proteins are made. But the microsomes are not in the nucleus, where the DNA and its instructions are. Something seems to be the matter. However, it is also known that little molecule pieces come off the DNA—not as long as the big DNA molecule that carries all the information itself, but like a small section of it. This is called RNA, but that is not essential. It is a kind of copy of the DNA, a short copy. The RNA, which somehow carries a message as to what kind of protein to make

The Relation of Physics to Other Sciences

goes over to the microsome; that is known. When it gets there, protein is synthesized at the microsome. That is also known. However, the details of how the amino acids come in and are arranged in accordance with a code that is on the RNA are, as yet, still unknown. We do not know how to read it. If we knew, for example, the "lineup" A, B, C, C, A, we could not tell you what protein is to be made.

Certainly no subject or field is making more progress on so many fronts at the present moment, than biology, and if we were to name the most powerful assumption of all, which leads one on and on in an attempt to understand life, it is that *all things are made of atoms*, and that everything that living things do can be understood in terms of the jigglings and wigglings of atoms.

Astronomy

In this rapid-fire explanation of the whole world, we must now turn to astronomy. Astronomy is older than physics. In fact, it got physics started by showing the beautiful simplicity of the motion of the stars and planets, the understanding of which was the *beginning* of physics. But the most remarkable discovery in all of astronomy is that *the stars are made of atoms of the same kind as those on the earth.** How was this done? Atoms liberate light which has definite

---

* How I'm rushing through this! How much each sentence in this brief story contains. "The stars are made of the same atoms as the earth." I usually pick one small topic like this to give a lecture on. Poets say science takes away from the beauty of the stars—mere globs of gas atoms. Nothing is "mere." I too can see the stars on a desert night, and feel them. But do I see less or more? The vastness of the heavens stretches my imagination—stuck on this carousel my little eye can catch one-million-year-old light. A vast pattern—of which I am a part—perhaps my stuff was belched from some forgotten star, as one is belching there. Or see them with the greater eye of Palomar, rushing all apart from some common starting point when they were perhaps all together. What is the pattern, or the meaning, or the

frequencies, something like the timbre of a musical instrument, which has definite pitches or frequencies of sound. When we are listening to several different tones we can tell them apart, but when we look with our eyes at a mixture of colors we cannot tell the parts from which it was made, because the eye is nowhere near as discerning as the ear in this connection. However, with a spectroscope we *can* analyze the frequencies of the light waves and in this way we can see the very tunes of the atoms that are in the different stars. As a matter of fact, two of the chemical elements were discovered on a star before they were discovered on the earth. Helium was discovered on the sun, whence its name, and technetium was discovered in certain cool stars. This, of course, permits us to make headway in understanding the stars, because they are made of the same kinds of atoms which are on the earth. Now we know a great deal about the atoms, especially concerning their behavior under conditions of high temperature but not very great density, so that we can analyze by statistical mechanics the behavior of the stellar substance. Even though we cannot reproduce the conditions on the earth, using the basic physical laws we often can tell precisely, or very closely, what will happen. So it is that physics aids astronomy. Strange as it may seem, we understand the distribution of matter in the interior of the sun far better than we understand the interior of the earth. What goes on *inside* a star is better understood than one might guess from the difficulty of having to look at a little dot of light through a telescope, because we can *calculate* what the atoms in the stars should do in most circumstances.

One of the most impressive discoveries was the origin of the energy of the stars, that makes them continue to burn. One of the

---

why? It does not do harm to the mystery to know a little about it. For far more marvelous is the truth than any artists of the past imagined! Why do the poets of the present not speak of it? What men are poets who can speak of Jupiter if he were like a man, but if he is an immense spinning sphere of methane and ammonia must be silent?

men who discovered this was out with his girl friend the night after he realized that *nuclear reactions* must be going on in the stars in order to make them shine. She said "Look at how pretty the stars shine!" He said, "Yes, and right now I am the only man in the world who knows *why* they shine." She merely laughed at him. She was not impressed with being out with the only man who, at that moment, knew why stars shine. Well, it is sad to be alone, but that is the way it is in this world.

It is the nuclear "burning" of hydrogen which supplies the energy of the sun; the hydrogen is converted into helium. Furthermore, ultimately, the manufacture of various chemical elements proceeds in the centers of the stars, from hydrogen. The stuff of which *we* are made, was "cooked" once, in a star, and spit out. How do we know? Because there is a clue. The proportion of the different isotopes—how much C^{12}, how much C^{13}, etc., is something which is never changed by *chemical* reactions, because the chemical reactions are so much the same for the two. The proportions are purely the result of *nuclear* reactions. By looking at the proportions of the isotopes in the cold, dead ember which we are, we can discover what the *furnace* was like in which the stuff of which we are made was formed. That furnace was like the stars, and so it is very likely that our elements were "made" in the stars and spit out in the explosions which we call novae and supernovae. Astronomy is so close to physics that we shall study many astronomical things as we go along.

Geology

We turn now to what are called *earth sciences*, or *geology*. First, meteorology and the weather. Of course the *instruments* of meteorology are physical instruments, and the development of experimental physics made these instruments possible, as was explained before. However, the theory of meteorology has never been satisfactorily worked out by the physicist. "Well," you say, "there is nothing but air, and we know the equations of the motions of air." Yes we do. "So if we know the condition of air today, why can't we figure out the condition of the air tomorrow?" First, we do not *really* know what the condition is today, because the air is swirling and twisting everywhere. It turns out to be very sensitive, and even unstable. If you have ever seen water run smoothly over a dam, and then turn into a large number of blobs and drops as it falls, you will understand what I mean by unstable. You know the condition of the water before it goes over the spillway; it is perfectly smooth; but the moment it begins to fall, where do the drops begin? What determines how big the lumps are going to be and where they will be? That is not known, because the water is unstable. Even a smooth moving mass of air going over a mountain turns into complex whirlpools and eddies. In many fields we find this situation of *turbulent flow* that we cannot analyze today. Quickly we leave the subject of weather, and discuss geology!

The question basic to geology is, what makes the earth the way it is? The most obvious processes are in front of your very eyes, the erosion processes of the rivers, the winds, etc. It is easy enough to understand these, but for every bit of erosion there is an equal amount of something else going on. Mountains are no lower today, on the average, than they were in the past. There must be mountain-*forming* processes. You will find, if you study geology, that there *are* mountain-forming processes and vulcanism, which nobody understands but which is half of geology. The phenomenon of volcanoes is really not understood. What makes an earthquake is, ultimately, not understood. It is understood that if something is pushing something else, it snaps and will slide— that is all right. But what pushes, and why? The theory is that there are currents inside the earth—circulating currents, due to the difference in temperature inside and outside—which, in their motion, push the surface slightly. Thus if there are two opposite circulations next to each other, the matter will collect in the region where they meet and make belts of mountains which are in unhappy stressed conditions, and so produce volcanoes and earth-quakes.

What about the inside of the earth? A great deal is known about the speed of earthquake waves through the earth and the density of distribution of the earth. However, physicists have been unable to get a good theory as to how dense a substance should be at the pressures that would be expected at the center of the earth. In other words, we cannot figure out the properties of matter very well in these circumstances. We do much less well with the earth than we do with the conditions of matter in the stars. The mathematics involved seems a little too difficult, so far, but perhaps it will not be too long before someone realizes that it is an important problem, and really work it out. The other aspect, of course, is that even if we did know the density, we cannot figure out the circulating currents. Nor can we really work out the properties of rocks at high pressure. We cannot tell how fast the rocks should "give"; that must all be worked out by experiment.

Psychology

Next, we consider the science of *psychology*. Incidentally, psychoanalysis is not a science: it is at best a medical process, and perhaps even more like witch-doctoring. It has a theory as to what causes disease—lots of different "spirits," etc. The witch doctor has a theory that a disease like malaria is caused by a spirit which comes into the air; it is not cured by shaking a snake over it, but quinine does help malaria. So, if you are sick, I would advise that you go to the witch doctor because he is the man in the tribe who knows the most about the disease; on the other hand, his knowledge is not science. Psychoanalysis has not been checked carefully by experiment, and there is no way to find a list of the number of cases in which it works, the number of cases in which it does not work, etc.

The other branches of psychology, which involve things like the physiology of sensation—what happens in the eye, and what happens in the brain—are, if you wish, less interesting. But some small but real progress has been made in studying them. One of the most interesting technical problems may or may not be called psychology. The central problem of the mind, if you will, or the nervous system, is this: when an animal learns something, it can do something different than it could before, and its brain cell must have changed too, if it is made out of atoms. *In what way is it different?* We do not know where to look, or what to look for, when something is memorized. We do not know what it means, or what change there is in the nervous system, when a fact is learned. This is a very important problem which has not been solved at all. Assuming, however, that there is some kind of memory thing, the brain is such an enormous mass of interconnecting wires and nerves that it probably cannot be analyzed in a straightforward manner. There is an analog of this to computing machines and computing elements, in that they also have a lot of lines, and they have some kind of element, analogous, perhaps, to the synapse, or connection of one nerve to another. This is a very interesting subject which we have not the time to discuss further—the relationship between thinking and computing machines. It must be appreciated, of course, that this subject will tell us very little about the real complexities of ordinary human behavior. All human beings are so different. It will be a long time before we get there. We must start much further back. If we could even figure out how a *dog* works, we would have gone pretty far. Dogs are easier to understand, but nobody yet knows how dogs work.

How did it get that way?

In order for physics to be useful to other sciences in a *theoretical* way, other than in the invention of instruments, the science in question must supply to the physicist a description of the object in

The Relation of Physics to Other Sciences

a physicist's language. They can say "why does a frog jump?," and the physicist cannot answer. If they tell him what a frog is, that there are so many molecules, there is a nerve here, etc., that is different. If they will tell us, more or less, what the earth or the stars are like, then we can figure it out. In order for physical theory to be of any use, we must know where the atoms are located. In order to understand the chemistry, we must know exactly what atoms are present, for otherwise we cannot analyze it. That is but one limitation, of course.

There is another *kind* of problem in the sister sciences which does not exist in physics; we might call it, for lack of a better term, the historical question. How did it get that way? If we understand all about biology, we will want to know how all the things which are on the earth got there. There is the theory of evolution, an important part of biology. In geology, we not only want to know how the mountains are forming, but how the entire earth was formed in the beginning, the origin of the solar system, etc. That, of course, leads us to want to know what kind of matter there was in the world. How did the stars evolve? What were the initial conditions? That is the problem of astronomical history. A great deal has been found out about the formation of stars, the formation of elements from which we were made, and even a little about the origin of the universe.

There is no historical question being studied in physics at the present time. We do not have a question, "Here are the laws of physics, how did they get that way?" We do not imagine, at the moment, that the laws of physics are somehow changing with time, that they were different in the past than they are at present. Of course they *may* be, and the moment we find they *are*, the historical question of physics will be wrapped up with the rest of the history of the universe, and then the physicist will be talking about the same problems as astronomers, geologists, and biologists.

Finally, there is a physical problem that is common to many fields, that is very old, and that has not been solved. It is not the problem of finding new fundamental particles, but something left over from a long time ago—over a hundred years. Nobody in physics has really been able to analyze it mathematically satisfactorily in spite of its importance to the sister sciences. It is the analysis of *circulating or turbulent fluids*. If we watch the evolution of a star, there comes a point where we can deduce that it is going to start convection, and thereafter we can no longer deduce what should happen. A few million years later the star explodes, but we cannot figure out the reason. We cannot analyze the weather. We do not know the patterns of motions that there should be inside the earth. The simplest form of the problem is to take a pipe that is very long and push water through it at high speed. We ask: to push a given amount of water through that pipe, how much pressure is needed? No one can analyze it from first principles and the properties of water. If the water flows very slowly, or if we use a thick goo like honey, then we can do it nicely. You will find that in your textbook. What we really cannot do is deal with actual, wet water running through a pipe. That is the central problem which we ought to solve some day, and we have not.

A poet once said, "The whole universe is in a glass of wine." We will probably never know in what sense he meant that, for poets do not write to be understood. But it is true that if we look at a glass of wine closely enough we see the entire universe. There are the things of physics: the twisting liquid which evaporates depending on the wind and weather, the reflections in the glass, and our imagination adds the atoms. The glass is a distillation of the earth's rocks, and in its composition we see the secrets of the universe's age, and the evolution of stars. What strange array of chemicals are in the wine? How did they come to be? There are the ferments, the enzymes, the substrates, and the products. There in wine is found the great generalization: all life is fermentation. Nobody can discover the chemistry of wine without discovering, as did Louis Pasteur, the cause of much disease. How vivid is the claret, pressing its existence into the consciousness that watches

The Relation of Physics to Other Sciences

it! If our small minds, for some convenience, divide this glass of wine, this universe, into parts—physics, biology, geology, astronomy, psychology, and so on—remember that nature does not know it! So let us put it all back together, not forgetting ultimately what it is for. Let it give us one more final pleasure: drink it and forget it all!

Four

CONSERVATION OF ENERGY

What is energy?

In this chapter, we begin our more detailed study of the different aspects of physics, having finished our description of things in general. To illustrate the ideas and the kind of reasoning that might be used in theoretical physics, we shall now examine one of the most basic laws of physics, the conservation of energy.

There is a fact, or if you wish, a *law*, governing all natural phenomena that are known to date. There is no known exception to this law—it is exact so far as we know. The law is called the *conservation of energy*. It states that there is a certain quantity, which we call energy, that does not change in the manifold changes which nature undergoes. That is a most abstract idea, because it is a mathematical principle; it says that there is a numerical quantity which does not change when something happens. It is not a description of a mechanism, or anything concrete; it is just a strange fact that we can calculate some number and when we finish watching nature go through her tricks and calculate the number again, it is the same. (Something like the bishop on a red square, and after a number of moves—details unknown—it is still on some red square.

It is a law of this nature.) Since it is an abstract idea, we shall illustrate the meaning of it by an analogy.

Imagine a child, perhaps "Dennis the Menace," who has blocks which are absolutely indestructible, and cannot be divided into pieces. Each is the same as the other. Let us suppose that he has 28 blocks. His mother puts him with his 28 blocks into a room at the beginning of the day. At the end of the day, being curious, she counts the blocks very carefully, and discovers a phenomenal law—no matter what he does with the blocks, there are always 28 remaining! This continues for a number of days, until one day there are only 27 blocks, but a little investigating shows that there is one under the rug—she must look everywhere to be sure that the number of blocks has not changed. One day, however, the number appears to change—there are only 26 blocks. Careful investigation indicates that the window was open, and upon looking outside, the other two blocks are found. Another day, careful count indicates that there are 30 blocks! This causes considerable consternation, until it is realized that Bruce came to visit, bringing his blocks with him, and he left a few at Dennis' house. After she has disposed of the extra blocks, she closes the window, does not let Bruce in, and then everything is going along all right, until one time she counts and finds only 25 blocks. However, there is a box in the room, a toy box, and the mother goes to open the toy box, but the boy says "No, do not open my toy box," and screams. Mother is not allowed to open the toy box. Being extremely curious, and somewhat ingenious, she invents a scheme! She knows that a block weighs three ounces, so she weighs the box at a time when she sees 28 blocks, and it weighs 16 ounces. The next time she wishes to check, she weighs the box again, subtracts sixteen ounces and divides by three. She discovers the following:

$$\begin{pmatrix} \text{number of} \\ \text{blocks seen} \end{pmatrix} + \frac{(\text{weight of box}) - 16 \text{ ounces}}{3 \text{ ounces}} = \text{constant.} \quad (4.1)$$

Conservation of Energy

There then appear to be some new deviations, but careful study indicates that the dirty water in the bathtub is changing its level. The child is throwing blocks into the water, and she cannot see them because it is so dirty, but she can find out how many blocks are in the water by adding another term to her formula. Since the original height of the water was 6 inches and each block raises the water a quarter of an inch, this new formula would be:

$$\begin{pmatrix} \text{number of} \\ \text{blocks seen} \end{pmatrix} + \frac{(\text{weight of box}) - 16 \text{ ounces}}{3 \text{ ounces}} + \frac{(\text{height of water}) - 6 \text{ inches}}{1/4 \text{ inch}} = \text{constant.} \quad (4.2)$$

In the gradual increase in the complexity of her world, she finds a whole series of terms representing ways of calculating how many blocks are in places where she is not allowed to look. As a result, she finds a complex formula, a quantity which *has to be computed*, which always stays the same in her situation.

What is the analogy of this to the conservation of energy? The most remarkable aspect that must be abstracted from this picture is that *there are no blocks.* Take away the first terms in (4.1) and (4.2) and we find ourselves calculating more or less abstract things. The analogy has the following points. First, when we are calculating the energy, sometimes some of it leaves the system and goes away, or sometimes some comes in. In order to verify the conservation of energy, we must be careful that we have not put any in or taken any out. Second, the energy has a large number of *different forms*, and there is a formula for each one. These are: gravitational energy, kinetic energy, heat energy, elastic energy, electrical energy, chemical energy, radiant energy, nuclear energy, mass energy. If we total up the formulas for each of these contributions, it will not change except for energy going in and out.

It is important to realize that in physics today, we have no knowledge of what energy *is*. We do not have a picture that energy comes in little blobs of a definite amount. It is not that way. However, there are formulas for calculating some numerical quantity, and when we add it all together it gives "28"—always the same number. It is an abstract thing in that it does not tell us the mechanism or the *reasons* for the various formulas.

Gravitational potential energy

Conservation of energy can be understood only if we have the formula for all of its forms. I wish to discuss the formula for gravitational energy near the surface of the Earth, and I wish to derive this formula in a way which has nothing to do with history but is simply a line of reasoning invented for this particular lecture to give you an illustration of the remarkable fact that a great deal about nature can be extracted from a few facts and close reasoning. It is an illustration of the kind of work theoretical physicists become involved in. It is patterned after a most excellent argument by Mr. Carnot on the efficiency of steam engines.*

Consider weight-lifting machines—machines which have the property that they lift one weight by lowering another. Let us also make a hypothesis: that *there is no such thing as perpetual motion* with these weight-lifting machines. (In fact, that there is no perpetual motion at all is a general statement of the law of conservation of energy.) We must be careful to define perpetual motion. First, let us do it for weight-lifting machines. If, when we have lifted and lowered a lot of weights and restored the machine to the original condition, we find that the net result is to have *lifted a weight*, then we have a perpetual motion machine because we can use that lifted weight to run something else. That is, *provided* the machine which lifted the weight is brought back to its exact *original condition*, and furthermore that it is completely *self-contained*—that it has not received the energy to lift that weight from some external source—like Bruce's blocks.

A very simple weight-lifting machine is shown in Fig. 4–1. This

---

* Our point here is not so much the result, (4.3), which in fact you may already know, as the possibility of arriving at it by theoretical reasoning.

Conservation of Energy

Figure 4–1. Simple weight-lifting machine.

machine lifts weights three units "strong." We place three units on one balance pan, and one unit on the other. However, in order to get it actually to work, we must lift a little weight off the left pan. On the other hand, we could lift a one-unit weight by lowering the three-unit weight, if we cheat a little by lifting a little weight off the other pan. Of course, we realize that with any *actual* lifting machine, we must add a little extra to get it to run. This we disregard, *temporarily*. Ideal machines, although they do not exist, do not require anything extra. A machine that we actually use can be, in a sense, *almost* reversible: that is, if it will lift the weight of three by lowering a weight of one, then it will also lift nearly the weight of one the same amount by lowering the weight of three.

We imagine that there are two classes of machines, those that are *not* reversible, which includes all real machines, and those that *are* reversible, which of course are actually not attainable no matter how careful we may be in our design of bearings, levers, etc. We suppose, however, that there is such a thing—a reversible machine—which lowers one unit of weight (a pound or any other unit) by one unit of distance, and at the same time lifts a three-unit weight. Call this reversible machine, Machine A. Suppose this particular reversible machine lifts the three-unit weight a distance X. Then suppose we have another machine, Machine B, which is not necessarily reversible, which also lowers a unit weight a unit distance, but which lifts three units a distance Y. We can now prove that Y is not higher than X; that is, it is impossible to build a machine that will lift a weight *any higher* than it will be lifted by a reversible machine. Let us see why. Let us suppose that Y were higher than X. We take a one-unit weight and lower it one unit height with Machine B, and that lifts the three-unit weight up a distance Y. Then we could lower the weight from Y to X, *obtaining free power*, and use the reversible Machine A, running backwards, to lower the three-unit weight a distance X and lift the one-unit weight by one unit height. This will put the one-unit weight back where it was before, and leave both machines ready to be used again! We would therefore have perpetual motion if Y were higher than X, which we assumed was impossible. With those assumptions, we thus deduce that Y *is not higher than* X, so that of all machines that can be designed, the reversible machine is the best.

We can also see that all reversible machines must lift to *exactly the same height*. Suppose that B were really reversible also. The argument that Y is not higher than X is, of course, just as good as it was before, but we can also make our argument the other way around, using the machines in the opposite order, and prove that X *is not higher than* Y. This, then, is a very remarkable observation because it permits us to analyze the height to which different machines are going to lift something *without looking at the interior mechanism*. We know at once that if somebody makes an enormously elaborate series of levers that lift three units a certain distance by lowering one unit by one unit distance, and we compare it with a simple lever which does the same thing and is fundamentally reversible, his machine will lift it no higher, but perhaps less high. If his machine is reversible, we also know exactly *how* high it will lift. To summarize: every reversible machine, no matter how it operates, which drops one pound one foot and lifts a three-pound weight always lifts it the same distance, X. This is clearly a universal law of great utility. The next question is, of course, what is X?

Suppose we have a reversible machine which is going to lift this distance X, three for one. We set up three balls in a rack which does not move, as shown in Fig. 4–2. One ball is held on a stage at a distance one foot above the ground. The machine can lift three balls, lowering one by a distance 1. Now, we have arranged that the platform which holds three balls has a floor and two shelves, exactly spaced at distance X, and further, that the rack which holds the balls is spaced at distance X, (a). First we roll the balls horizontally from

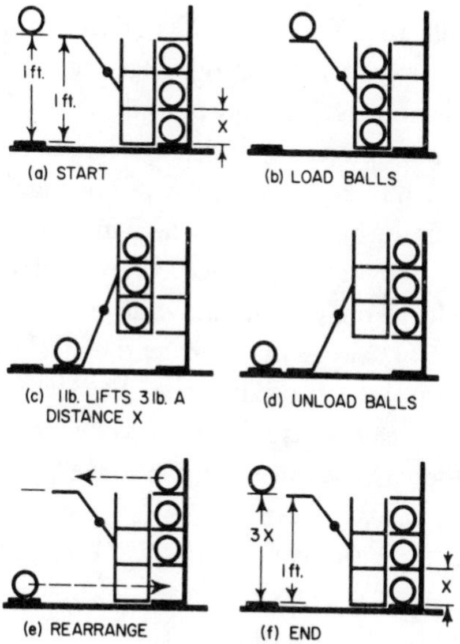

Conservation of Energy

Figure 4–2. A reversible machine.

the rack to the shelves, (b), and we suppose that this takes no energy because we do not change the height. The reversible machine then operates: it lowers the single ball to the floor, and it lifts the rack a distance X, (c). Now we have ingeniously arranged the rack so that these balls are again even with the platforms. Thus we unload the balls onto the rack, (d); having unloaded the balls, we can restore the machine to its original condition. Now we have three balls on the upper three shelves and one at the bottom. But the strange thing is that, in a certain way of speaking, we have not lifted *two* of them at all because, after all, there were balls on shelves 2 and 3 before. The resulting effect has been to lift *one ball* a distance $3X$. Now, if $3X$ exceeds one foot, then we can *lower* the ball to return the machine to the initial condition, (f), and we can run the apparatus again. Therefore $3X$ cannot exceed one foot, for if $3X$ exceeds one foot we can make perpetual motion. Likewise, we can prove that *one foot cannot exceed* $3X$, by making the whole machine run the opposite way, since it is a reversible machine. Therefore $3X$ is neither *greater nor less than a foot*, and we discover then, by argument alone, the law that $X = \frac{1}{3}$ foot. The generalization is clear: one pound falls a certain distance in operating a reversible machine; then the machine can lift p pounds this distance divided by p. Another way of putting the result is that three pounds times the height lifted, which in our problem was X, is equal to one pound times the distance lowered, which is one foot in this case. If we take all the weights and multiply them by the heights at which they are now, above the floor, let the machine operate, and then multiply all the weights by all the heights again, *there will be no change.* (We have to generalize the example where we moved only one weight to the case where when we lower one we lift several different ones—but that is easy.)

We call the sum of the weights times the heights *gravitational potential energy*—the energy which an object has because of its relationship in space, relative to the earth. The formula for gravitational energy, then, so long as we are not too far from the earth (the force weakens as we go higher) is

$$\begin{pmatrix} \text{gravitational} \\ \text{potential energy} \\ \text{for one object} \end{pmatrix} \quad = \quad (\text{weight}) \times (\text{height}). \qquad (4.3)$$

It is a very beautiful line of reasoning. The only problem is that perhaps it is not true. (After all, nature does not *have* to go along with our reasoning.) For example, perhaps perpetual motion is, in fact, possible. Some of the assumptions may be wrong, or we may have made a mistake in reasoning, so it is always necessary to check. *It turns out experimentally*, in fact, to be true.

The general name of energy which has to do with location relative to something else is called *potential* energy. In this particular case, of course, we call it *gravitational potential energy*. If it is a question of electrical forces against which we are working, instead of gravitational forces, if we are "lifting" charges away from other charges with a lot of levers, then the energy content is called *electrical*

Conservation of Energy

potential energy. The general principle is that the change in the energy is the force times the distance that the force is pushed, and that this is a change in energy in general:

$$\begin{pmatrix} \text{change in} \\ \text{energy} \end{pmatrix} = (\text{force}) \times \begin{pmatrix} \text{distance force} \\ \text{acts through} \end{pmatrix} \tag{4.4}$$

We will return to many of these other kinds of energy as we continue the course.

The principle of the conservation of energy is very useful for deducing what will happen in a number of circumstances. In high school we learned a lot of laws about pulleys and levers used in different ways. We can now see that these "laws" are *all the same thing*, and that we did not have to memorize 75 rules to figure it out. A simple example is a smooth inclined plane which is, happily, a three-four-five triangle (Fig. 4–3). We hang a one-pound weight on the inclined plane with a pulley, and on the other side of the pulley, a weight W. We want to know how heavy W must be to balance the one pound on the plane. How can we figure that out? If we say it is just balanced, it is reversible and so can move up and down, and we can consider the following situation. In the initial circumstance, (a), the one pound weight is at the bottom and weight W is at the top. When W has slipped down in a reversible way, we have a one-pound weight at the top and the weight W the slant distance, (b), or five feet, from the plane in which it was before. We *lifted* the one-pound weight only *three* feet and we lowered W pounds by *five* feet.

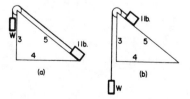

Figure 4–3. Inclined plane.

Therefore $W = \frac{3}{5}$ of a pound. Note that we deduced this from the *conservation of energy*, and not from force components. Cleverness, however, is relative. It can be deduced in a way which is even more brilliant, discovered by Stevinus and inscribed on his tombstone. Figure 4–4 explains that it has to be $\frac{3}{5}$ of a pound, because the chain does not go around. It is evident that the lower part of the chain is balanced by itself, so that the pull of the five weights on one side must balance the pull of three weights on the other, or whatever the ratio of the legs. You see, by looking at this diagram, that W must be $\frac{3}{5}$ of a pound. (If you get an epitaph like that on your gravestone, you are doing fine.)

Let us now illustrate the energy principle with a more complicated problem, the screw jack shown in Fig. 4–5. A handle 20 inches long is used to turn the screw, which has 10 threads to the inch. We would like to know how much force would be needed at the handle to lift one ton (2000 pounds). If we want to lift the ton one inch, say, then we must turn the handle around ten times. When it goes around once it goes approximately 126 inches. The handle must thus travel 1260 inches, and if we used various pulleys, etc., we would be lifting our one ton with an unknown smaller weight W applied to the end of the handle. So we find out that W is about 1.6 pounds. This is a result of the conservation of energy.

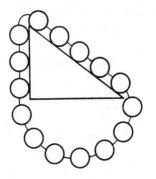

Figure 4–4. The epitaph of Stevinus.

Conservation of Energy

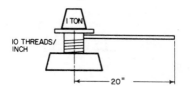

Figure 4–5. A screw jack.

Take now the somewhat more complicated example shown in Fig. 4–6. A rod or bar, 8 feet long, is supported at one end. In the middle of the bar is a weight of 60 pounds, and at a distance of two feet from the support there is a weight of 100 pounds. How hard do we have to lift the end of the bar in order to keep it balanced, disregarding the weight of the bar? Suppose we put a pulley at one end and hang a weight on the pulley. How big would the weight W have to be in order for it to balance? We imagine that the weight falls any arbitrary distance—to make it easy for ourselves suppose it goes down 4 inches—how high would the two load weights rise? The center rises 2 inches, and the point a quarter of the way from the fixed end lifts 1 inch. Therefore, the principle that the sum of the heights times the weights does not change tells us that the weight W times 4 inches down, plus 60 pounds times 2 inches up, plus 100 pounds times 1 inch has to add up to nothing:

$$-4W + (2)(60) + (1)(100) = 0, \qquad W = 55 \text{ lb.} \qquad (4.5)$$

Thus we must have a 55-pound weight to balance the bar. In this way we can work out the laws of "balance"—the statics of complicated bridge arrangements, and so on. This approach is called the *principle of virtual work*, because in order to apply this argument we had to *imagine* that the structure moves a little—even though it is

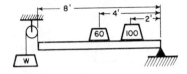

Figure 4–6. Weighted rod supported on one end.

not *really* moving or even *movable*. We use the very small imagined motion to apply the principle of conservation of energy.

Kinetic energy

To illustrate another type of energy we consider a pendulum (Fig. 4–7). If we pull the mass aside and release it, it swings back and forth. In its motion, it loses height in going from either end to the center. Where does the potential energy go? Gravitational energy disappears when it is down at the bottom; nevertheless, it will climb up again. The gravitational energy must have gone into another form. Evidently it is by virtue of its *motion* that it is able to climb up again, so we have the conversion of gravitational energy into some other form when it reaches the bottom.

We must get a formula for the energy of motion. Now, recalling our arguments about reversible machines, we can easily see that in the motion at the bottom must be a quantity of energy which permits it to rise a certain height, and which has nothing to do with the *machinery* by which it comes up or the *path* by which it comes up. So we have an equivalence formula something like the one we wrote for the child's blocks. We have another form to represent the energy. It is easy to say what it is. The kinetic energy at the bottom equals the weight times the height that it could go, corresponding to its velocity: K.E. = *WH*. What we need is the formula which tells us the height by some rule that has to do with the motion of objects. If we start something out with a certain

Conservation of Energy

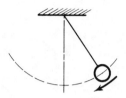

Figure 4–7. Pendulum.

velocity, say straight up, it will reach a certain height; we do not know what it is yet, but it depends on the velocity—there is a formula for that. Then to find the formula for kinetic energy for an object moving with velocity V, we must calculate the height that it could reach, and multiply by the weight. We shall soon find that we can write it this way:

$$\text{K.E.} = WV^2/2g. \tag{4.6}$$

Of course, the fact that motion has energy has nothing to do with the fact that we are in a gravitational field. It makes no difference *where* the motion came from. This is a general formula for various velocities. Both (4.3) and (4.6) are approximate formulas, the first because it is incorrect when the heights are great, i.e., when the heights are so high that gravity is weakening; the second, because of the relativistic correction at high speeds. However, when we do finally get the exact formula for the energy, then the law of conservation of energy is correct.

Other forms of energy

We can continue in this way to illustrate the existence of energy in other forms. First, consider elastic energy. If we pull down on a spring, we must do some work, for when we have it down, we can lift weights with it. Therefore in its stretched condition it has a possibility of doing some work. If we were to evaluate the sums of weights times heights, it would not check out—we must add something else to account for the fact that the spring is under tension. Elastic energy is the formula for a spring when it is stretched. How much energy is it? If we let go, the elastic energy, as the spring passes through the equilibrium point, is converted to kinetic energy and it goes back and forth between compressing or stretching the spring and kinetic energy of motion. (There is also some gravitational energy going in and out, but we can do this experiment "sideways" if we like.) It keeps going until the losses—Aha! We have cheated all the way through by putting on little weights to move things or saying that the machines are reversible, or that they go on forever, but we can see that things do stop, eventually. Where is the energy when the spring has finished moving up and down? This brings in *another* form of energy: *heat energy*.

Inside a spring or a lever there are crystals which are made up of lots of atoms, and with great care and delicacy in the arrangement of the parts one can try to adjust things so that as something rolls on something else, none of the atoms do any jiggling at all. But one must be very careful. Ordinarily when things roll, there is bumping and jiggling because of the irregularities of the material, and the atoms start to wiggle inside. So we lose track of that energy; we find the atoms are wiggling inside in a random and confused manner after the motion slows down. There is still kinetic energy, all right, but it is not associated with visible motion. What a dream! How do we *know* there is still kinetic energy? It turns out that with thermometers you can find out that, in fact, the spring or the lever is *warmer*, and that there is really an increase of kinetic energy by a definite amount. We call this form of energy *heat energy*, but we know that it is not really a new form, it is just kinetic energy—internal motion. (One of the difficulties with all these experiments with matter that we do on a large scale is that we cannot really demonstrate the conservation of energy and we cannot really make our reversible machines, because every time we move a large clump of stuff, the atoms do not remain absolutely undisturbed, and so a

Conservation of Energy

certain amount of random motion goes into the atomic system. We cannot see it, but we can measure it with thermometers, etc.)

There are many other forms of energy, and of course we cannot describe them in any more detail just now. There is electrical energy, which has to do with pushing and pulling by electric charges. There is radiant energy, the energy of light, which we know is a form of electrical energy because light can be represented as wigglings in the electromagnetic field. There is chemical energy, the energy which is released in chemical reactions. Actually, elastic energy is, to a certain extent, like chemical energy, because chemical energy is the energy of the attraction of the atoms, one for the other, and so is elastic energy. Our modern understanding is the following: chemical energy has two parts, kinetic energy of the electrons inside the atoms, so part of it is kinetic, and electrical energy of interaction of the electrons and the protons—the rest of it, therefore, is electrical. Next we come to nuclear energy, the energy which is involved with the arrangement of particles inside the nucleus, and we have formulas for that, but we do not have the fundamental laws. We know that it is not electrical, not gravitational, and not purely chemical, but we do not know what it is. It seems to be an additional form of energy. Finally, associated with the relativity theory, there is a modification of the laws of kinetic energy, or whatever you wish to call it, so that kinetic energy is combined with another thing called *mass energy*. An object has energy from its sheer *existence*. If I have a positron and an electron, standing still doing nothing—never mind gravity, never mind anything—and they come together and disappear, radiant energy will be liberated, in a definite amount, and the amount can be calculated. All we need know is the mass of the object. It does not depend on what it is—we make two things disappear, and we get a certain amount of energy. The formula was first found by Einstein; it is $E = mc^2$.

It is obvious from our discussion that the law of conservation of energy is enormously useful in making analyses, as we have illustrated in a few examples without knowing all the formulas. If we had all the formulas for all kinds of energy, we could analyze how many processes should work without having to go into the details. Therefore conservation laws are very interesting. The question naturally arises as to what other conservation laws there are in physics. There are two other conservation laws which are analogous to the conservation of energy. One is called the conservation of linear momentum. The other is called the conservation of angular momentum. We will find out more about these later. In the last analysis, we do not understand the conservation laws deeply. We do not understand the conservation of energy. We do not understand energy as a certain number of little blobs. You may have heard that photons come out in blobs and that the energy of a photon is Planck's constant times the frequency. That is true, but since the frequency of light can be anything, there is no law that says that energy has to be a certain definite amount. Unlike Dennis' blocks, there can be any amount of energy, at least as presently understood. So we do not understand this energy as counting something at the moment, but just as a mathematical quantity, which is an abstract and rather peculiar circumstance. In quantum mechanics it turns out that the conservation of energy is very closely related to another important property of the world, *things do not depend on the absolute time.* We can set up an experiment at a given moment and try it out, and then do the same experiment at a later moment, and it will behave in exactly the same way. Whether this is strictly true or not, we do not know. If we assume that it *is* true, and add the principles of quantum mechanics, then we can deduce the principle of the conservation of energy. It is a rather subtle and interesting thing, and it is not easy to explain. The other conservation laws are also linked together. The conservation of momentum is associated in quantum mechanics with the proposition that it makes no difference *where* you do the experiment, the results will always be the same. As independence in space has to do with the conservation of momentum, independence of time has to do with the conservation of energy, and finally, if we *turn* our apparatus, this too makes no difference, and so the invariance of the world to angular orientation

Conservation of Energy

is related to the conservation of *angular momentum*. Besides these, there are three other conservation laws, that are exact so far as we can tell today, which *are* much simpler to understand because they are in the nature of counting blocks.

The first of the three is the *conservation of charge*, and that merely means that you count how many positive, minus how many negative electrical charges you have, and the number is never changed. You may get rid of a positive with a negative, but you do not create any net excess of positives over negatives. Two other laws are analogous to this one—one is called the *conservation of baryons*. There are a number of strange particles, a neutron and a proton are examples, which are called baryons. In any reaction whatever in nature, if we count how many baryons are coming into a process, the number of baryons* which come out will be exactly the same. There is another law, the *conservation of leptons*. We can say that the group of particles called leptons are: electron, mu meson, and neutrino. There is an antielectron which is a positron, that is, a −1 lepton. Counting the total number of leptons in a reaction reveals that the number in and out never changes, at least so far as we know at present.

These are the six conservation laws, three of them subtle, involving space and time, and three of them simple, in the sense of counting something.

With regard to the conservation of energy, we should note that *available* energy is another matter—there is a lot of jiggling around in the atoms of the water of the sea, because the sea has a certain temperature, but it is impossible to get them herded into a definite motion without taking energy from somewhere else. That is, although we know for a fact that energy is conserved, the energy available for human utility is not conserved so easily. The laws which govern how much energy is available are called the *laws of thermodynamics* and involve a concept called entropy for irreversible thermodynamic processes.

---

* Counting antibaryons as −1 baryon.

Finally, we remark on the question of where we can get our supplies of energy today. Our supplies of energy are from the sun, rain, coal, uranium, and hydrogen. The sun makes the rain, and the coal also, so that all these are from the sun. Although energy is conserved, nature does not seem to be interested in it; she liberates a lot of energy from the sun, but only one part in two billion falls on the earth. Nature has conservation of energy, but does not really care; she spends a lot of it in all directions. We have already obtained energy from uranium; we can also get energy from hydrogen, but at present only in an explosive and dangerous condition. If it can be controlled in thermonuclear reactions, it turns out that the energy that can be obtained from 10 quarts of water per second is equal to all of the electrical power generated in the United States. With 150 gallons of running water a minute, you have enough fuel to supply all the energy which is used in the United States today! Therefore it is up to the physicist to figure out how to liberate us from the need for having energy. It can be done.

Five

THE THEORY OF GRAVITATION

Planetary motions

In this chapter we shall discuss one of the most far-reaching generalizations of the human mind. While we are admiring the human mind, we should take some time off to stand in awe of a *nature* that could follow with such completeness and generality such an elegantly simple principle as the law of gravitation. What is this law of gravitation? It is that every object in the universe attracts every other object with a force which for any two bodies is proportional to the mass of each and varies inversely as the square of the distance between them. This statement can be expressed mathematically by the equation

$$F = G\frac{mm'}{r^2}.$$

If to this we add the fact that an object responds to a force by accelerating in the direction of the force by an amount that is inversely proportional to the mass of the object, we shall have said everything required, for a sufficiently talented mathematician could then deduce all the consequences of these two principles. However, since you are not assumed to be sufficiently talented yet, we shall discuss the consequences in more detail, and not just leave you with only these two bare principles. We shall briefly relate the story of the discovery of the law of gravitation and discuss some of its consequences, its effects on history, the mysteries that such a law entails, and some refinements of the law made by Einstein; we shall also discuss the relationships of the law to the other laws of physics. All this cannot be done in one chapter, but these subjects will be treated in due time in subsequent chapters.

The story begins with the ancients observing the motions of planets among the stars, and finally deducing that they went around the sun, a fact that was rediscovered later by Copernicus. Exactly *how* the planets went around the sun, with exactly *what motion*, took a little more work to discover. In the beginning of the fifteenth century there were great debates as to whether they really went around the sun or not. Tycho Brahe had an idea that was different from anything proposed by the ancients: his idea was that these debates about the nature of the motions of the planets would best be resolved if the actual positions of the planets in the sky were measured sufficiently accurately. If measurement showed exactly how the planets moved, then perhaps it would be possible to establish one or another viewpoint. This was a tremendous idea—that to find something out, it is better to perform some careful experiments than to carry on deep philosophical arguments. Pursuing this idea, Tycho Brahe studied the positions of the planets for many years in his observatory on the island of Hven, near Copenhagen. He made voluminous tables, which were then studied by the mathematician Kepler, after Tycho's death. Kepler discovered from the data some very beautiful and remarkable, but simple, laws regarding planetary motion.

Kepler's laws

First of all, Kepler found that each planet goes around the sun in a curve called an *ellipse*, with the sun at a focus of the ellipse. An ellipse is not just an oval, but is a very specific and precise curve that

The Theory of Gravitation

can be obtained by using two tacks, one at each focus, a loop of string, and a pencil; more mathematically, it is the locus of all points the sum of whose distances from two fixed points (the foci) is a constant. Or, if you will, it is a foreshortened circle (Fig. 5–1).

Kepler's second observation was that the planets do not go around the sun at a uniform speed, but move faster when they are nearer the sun and more slowly when they are farther from the sun, in precisely this way: Suppose a planet is observed at any two successive times, let us say a week apart, and that the radius vector* is drawn to the planet for each observed position. The orbital arc traversed by the planet during the week, and the two radius vectors, bound a certain plane area, the shaded area shown in Fig. 5–2. If two similar observations are made a week apart, at a part of the orbit farther from the sun (where the planet moves more slowly), the similarly bounded area is exactly the same as in the first case. So, in accordance with the second law, the orbital speed of each planet is such that the radius "sweeps out" equal areas in equal times.

Finally, a third law was discovered by Kepler much later; this law is of a different category from the other two, because it deals not with only a single planet, but relates one planet to another. This law says that when the orbital period and orbit size of any two planets are compared, the periods are proportional to the ½ power of the

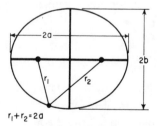

Figure 5–1. An ellipse.

---

* A radius vector is a line drawn from the sun to any point in a planet's orbit.

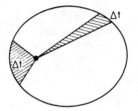

Figure 5–2. Kepler's law of areas.

orbit size. In this statement the period is the time interval it takes a planet to go completely around its orbit, and the size is measured by the length of the greatest diameter of the elliptical orbit, technically known as the major axis. More simply, if the planets went in circles, as they nearly do, the time required to go around the circle would be proportional to the ³⁄₂ power of the diameter (or radius). Thus Kepler's three laws are:

I. Each planet moves around the sun in an ellipse, with the sun at one focus.

II. The radius vector from the sun to the planet sweeps out equal areas in equal intervals of time.

III. The squares of the periods of any two planets are proportional to the cubes of the semimajor axes of their respective orbits: $T \sim a^{3/2}$.

Development of dynamics

While Kepler was discovering these laws, Galileo was studying the laws of motion. The problem was, what makes the planets go around? (In those days, one of the theories proposed was that the planets went around because behind them were invisible angels, beating their wings and driving the planets forward. You will see that this theory is now modified! It turns out that in order to keep the planets going around, the invisible angels must fly in a different

The Theory of Gravitation

direction and they have no wings. Otherwise, it is a somewhat similar theory!) Galileo discovered a very remarkable fact about motion, which was essential for understanding these laws. That is the principle of *inertia*—if something is moving, with nothing touching it and completely undisturbed, it will go on forever, coasting at a uniform speed in a straight line. (*Why* does it keep on coasting? We do not know, but that is the way it is.)

Newton modified this idea, saying that the only way to change the motion of a body is to use *force*. If the body speeds up, a force has been applied *in the direction of motion*. On the other hand, if its motion is changed to a new *direction*, a force has been applied *sideways*. Newton thus added the idea that a force is needed to change the speed *or the direction* of motion of a body. For example, if a stone is attached to a string and is whirling around in a circle, it takes a force to keep it in the circle. We have to *pull* on the string. In fact, the law is that the acceleration produced by the force is inversely proportional to the mass, or the force is proportional to the mass times the acceleration. The more massive a thing is, the stronger the force required to produce a given acceleration. (The mass can be measured by putting other stones on the end of the same string and making them go around the same circle at the same speed. In this way it is found that more or less force is required, the more massive object requiring more force.) The brilliant idea resulting from these considerations is that no *tangential* force is needed to keep a planet in its orbit (the angels do not have to fly tangentially) because the planet would coast in that direction anyway. If there were nothing at all to disturb it, the planet would go off in a *straight line*. But the actual motion deviates from the line on which the body would have gone if there were no force, the deviation being essentially *at right angles* to the motion, not in the direction of the motion. In other words, because of the principle of inertia, the force needed to control the motion of a planet around the sun is not a force *around* the sun but *toward* the sun. (If there is a force toward the sun, the sun might be the angel, of course!)

Newton's law of gravitation

From his better understanding of the theory of motion, Newton appreciated that *the sun* could be the seat or organization of forces that govern the motion of the planets. Newton proved to himself (and perhaps we shall be able to prove it soon) that the very fact that equal areas are swept out in equal times is a precise sign post of the proposition that all deviations are precisely *radial*—that the law of areas is a direct consequence of the idea that all of the forces are directed exactly *toward the sun*.

Next, by analyzing Kepler's third law it is possible to show that the farther away the planet, the weaker the forces. If two planets at different distances from the sun are compared, the analysis shows that the forces are inversely proportional to the squares of the respective distances. With the combination of the two laws, Newton concluded that there must be a force, inversely as the square of the distance, directed in a line between the two objects.

Being a man of considerable feeling for generalities, Newton supposed, of course, that this relationship applied more generally than just to the sun holding the planets. It was already known, for example, that the planet Jupiter had moons going around it as the moon of the earth goes around the earth, and Newton felt certain that each planet held its moons with a force. He already knew of the force holding *us* on the earth, so he proposed that this was a *universal* force—*that everything pulls everything else.*

The next problem was whether the pull of the earth on its people was the "same" as its pull on the moon, i.e., inversely as the square of the distance. If an object on the surface of the earth falls 16 feet in the first second after it is released from rest, how far does the moon fall in the same time? We might say that the moon does not fall at all. But if there were no force on the moon, it would go off in a straight line, whereas it goes in a circle instead, so it really *falls in* from where it would have been if there were no force at all. We can calculate from the radius of the moon's orbit (which is about 240,000 miles) and how long it takes to go around the earth (ap-

The Theory of Gravitation

proximately 29 days), how far the moon moves in its orbit in 1 second, and can then calculate how far it falls in one second.* This distance turns out to be roughly $\frac{1}{20}$ of an inch in a second. That fits very well with the inverse square law, because the earth's radius is 4000 miles, and if something which is 4000 miles from the center of the earth falls 16 feet in a second, something 240,000 miles, or 60 times as far away, should fall only $\frac{1}{3600}$ of 16 feet, which also is roughly $\frac{1}{20}$ of an inch. Wishing to put this theory of gravitation to a test by similar calculations, Newton made his calculations very carefully and found a discrepancy so large that he regarded the theory as contradicted by facts, and did not publish his results. Six years later a new measurement of the size of the earth showed that the astronomers had been using an incorrect distance to the moon. When Newton heard of this, he made the calculation again, with the corrected figures, and obtained beautiful agreement.

This idea that the moon "falls" is somewhat confusing, because, as you see, it does not come any *closer*. The idea is sufficiently interesting to merit further explanation: the moon falls in the sense that *it falls away from the straight line that it would pursue if there were no forces*. Let us take an example on the surface of the earth. An object released near the earth's surface will fall 16 feet in the first second. An object shot out *horizontally* will also fall 16 feet; even though it is moving horizontally, it still falls the same 16 feet in the same time. Figure 5–3 shows an apparatus which demonstrates this. On the horizontal track is a ball which is going to be driven forward a little distance away. At the same height is a ball which is going to fall vertically, and there is an electrical switch arranged so that at the moment the first ball leaves the track, the second ball is released. That they come to the same depth at the same time is witnessed by the fact that they collide in midair. An object like a bullet, shot horizontally, might go a long way in one second—perhaps 2000

---

* That is, how far the circle of the moon's orbit falls below the straight line tangent to it at the point where the moon was one second before.

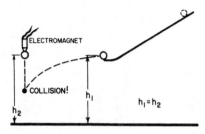

Figure 5–3. Apparatus for showing the independence of vertical and horizontal motions.

feet—but it will still fall 16 feet if it is aimed horizontally. What happens if we shoot a bullet faster and faster? Do not forget that the earth's surface is curved. If we shoot it fast enough, then when it falls 16 feet it may be at just the same height above the ground as it was before. How can that be? It still falls, but the earth curves away, so it falls "around" the earth. The question is, how far does it have to go in one second so that the earth is 16 feet below the horizon? In Fig. 5–4 we see the earth with its 4000-mile radius, and the tangential, straightline path that the bullet would take if there were no force. Now, if we use one of those wonderful theorems in geometry, which says that our tangent is the mean proportional between the two parts of the diameter cut by an equal chord, we see that the horizontal distance travelled is the mean proportional between the 16 feet fallen and the 8000-mile diameter of the earth. The square root of $(16/5280) \times 8000$ comes out very close to 5 miles. Thus we see that if the bullet moves at 5 miles a second, it then will continue to fall toward the earth at the same rate of 16 feet each second, but will never get any closer because the earth keeps curving away from it. Thus it was that Mr. Gagarin maintained himself in space while going 25,000 miles around the earth at approximately 5 miles per second. (He took a little longer because he was a little higher.)

The Theory of Gravitation

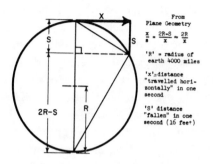

Figure 5–4. Acceleration toward the center of a circular path. From plane geometry, $x/s = (2R - S)/x \approx 2R/x$, where R is the radius of the earth, 4000 miles; x is the distance "travelled horizontally" in one second; and S is the distance "fallen" in one second (16 feet).

Any great discovery of a new law is useful only if we can take more out than we put in. Now, Newton *used* the second and third of Kepler's laws to deduce his law of gravitation. What did he *predict?* First, his analysis of the moon's motion was a prediction because it connected the falling of objects on the earth's surface with that of the moon. Second, the question is, *is the orbit an ellipse?* We shall see in a later chapter how it is possible to calculate the motion exactly, and indeed one can prove that it should be an ellipse,* so no extra fact is needed to explain Kepler's *first* law. Thus Newton made his first powerful prediction.

The law of gravitation explains many phenomena not previously understood. For example, the pull of the moon on the earth causes the tides, hitherto mysterious. The moon pulls the water up under it and makes the tides—people had thought of that before, but they were not as clever as Newton, and so they thought there ought to be only one tide during the day. The reasoning was that the moon pulls the water up under it, making a high tide and a low tide, and since

---

* The proof is not given in this course.

the earth spins underneath, that makes the tide at one station go up and down every 24 hours. Actually the tide goes up and down in 12 hours. Another school of thought claimed that the high tide should be on the other side of the earth because, so they argued, the moon pulls the earth away from the water! Both of these theories are wrong. It actually works like this: the pull of the moon for the earth and for the water is "balanced" at the center. But the water which is closer to the moon is pulled *more* than the average and the water which is farther away from it is pulled *less* than the average. Furthermore, the water can flow while the more rigid earth cannot. The true picture is a combination of these two things.

What do we mean by "balanced"? What balances? If the moon pulls the whole earth toward it, why doesn't the earth fall right "up" to the moon? Because the earth does the same trick as the moon, it goes in a circle around a point which is inside the earth but not at its center. The moon does not just go around the earth, the earth and the moon both go around a central position, each falling toward this common position, as shown in Fig. 5–5. This motion around the common center is what balances the fall of each. So the earth is not going in a straight line either; it travels in a circle. The water on the far side is "unbalanced" because the moon's attraction there is weaker than it is at the center of the earth, where it just balances the "centrifugal force." The result of this imbalance is that the water rises up, away from the center of the earth. On the near side, the attraction from the moon is stronger, and the imbalance is in the opposite direction in space, but again *away* from the center of the earth. The net result is that we get *two* tidal bulges.

Universal gravitation

What else can we understand when we understand gravity? Everyone knows the earth is round. Why is the earth round? That is easy; it is due to gravitation. The earth can be understood to be round merely because everything attracts everything else and so it has

The Theory of Gravitation

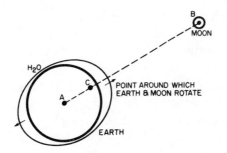

Figure 5–5. The earth-moon system, with tides.

attracted itself together as far as it can! If we go even further, the earth is not *exactly* a sphere because it is rotating, and this brings in centrifugal effects which tend to oppose gravity near the equator. It turns out that the earth should be elliptical, and we even get the right shape for the ellipse. We can thus deduce that the sun, the moon, and the earth should be (nearly) spheres, just from the law of gravitation.

What else can you do with the law of gravitation? If we look at the moons of Jupiter we can understand everything about the way they move around that planet. Incidentally, there was once a certain difficulty with the moons of Jupiter that is worth remarking on. These satellites were studied very carefully by Roemer, who noticed that the moons sometimes seemed to be ahead of schedule, and sometimes behind. (One can find their schedules by waiting a very long time and finding out how long it takes on the average for the moons to go around.) Now they were *ahead* when Jupiter was particularly *close* to the earth and they were *behind* when Jupiter was *farther* from the earth. This would have been a very difficult thing to explain according to the law of gravitation—it would have been, in fact, the death of this wonderful theory if there were no other explanation. If a law does not work even in *one place* where it ought to, it is just wrong. But the reason for this discrepancy was very simple and beautiful: it takes a little while to *see* the moons of Jupiter because of the time it takes light to travel from Jupiter to the earth.

When Jupiter is closer to the earth the time is a little less, and when it is farther from the earth, the time is more. This is why moons appear to be, on the average, a little ahead or a little behind, depending on whether they are closer to or farther from the earth. This phenomenon showed that light does not travel instantaneously, and furnished the first estimate of the speed of light. This was done in 1656.

If all of the planets push and pull on each other, the force which controls, let us say, Jupiter in going around the sun is not just the force from the sun; there is also a pull from, say, Saturn. This force is not really strong, since the sun is much more massive than Saturn, but there is *some* pull, so the orbit of Jupiter should not be a perfect ellipse, and it is not; it is slightly off, and "wobbles" around the correct elliptical orbit. Such a motion is a little more complicated. Attempts were made to analyze the motions of Jupiter, Saturn, and Uranus on the basis of the law of gravitation. The effects of each of these planets on each other were calculated to see whether or not the tiny deviations and irregularities in these motions could be completely understood from this one law. Lo and behold, for Jupiter and Saturn, all was well, but Uranus was "weird." It behaved in a very peculiar manner. It was not travelling in an exact ellipse, but that was understandable, because of the attractions of Jupiter and Saturn. But even if allowance were made for these attractions, Uranus *still* was not going right, so the laws of gravitation were in danger of being overturned, a possibility that could not be ruled out. Two men, Adams and Leverrier, in England and France, independently, arrived at another possibility: perhaps there is *another* planet, dark and invisible, which men had not seen. This planet, N, could pull on Uranus. They calculated where such a planet would have to be in order to cause the observed perturbations. They sent messages to the respective observatories, saying, "Gentlemen, point your telescope to such and such a place, and you will see a new planet." It often depends on with whom you are working as to whether they pay any attention to you or not. They did pay attention to Leverrier; they looked, and there planet N was! The other observatory then also looked very quickly in the next few days and saw it too.

The Theory of Gravitation

This discovery shows that Newton's laws are absolutely right in the solar system; but do they extend beyond the relatively small distances of the nearest planets? The first test lies in the question, do *stars* attract *each other* as well as planets? We have definite evidence that they do in the *double stars*. Figure 5–6 shows a double star—two stars very close together (there is also a third star in the picture so that we will know that the photograph was not turned). The stars are also shown as they appeared several years later. We see that, relative to the "fixed" star, the axis of the pair has rotated, i.e., the two stars are going around each other. Do they rotate according to Newton's laws? Careful measurements of the relative positions of one such double star system are shown in Fig. 5–7. There we see a beautiful ellipse, the measures starting in 1862 and going all the way around to 1904 (by now it must have gone around once more). Everything coincides with Newton's laws, except that the star Sirius A is *not at the focus*. Why should that be? Because the plane of the ellipse is not in the "plane of the sky." We are not looking at right angles to the orbit plane, and when an ellipse is viewed at a tilt, it remains an ellipse but the focus is no longer at the same place. Thus we can analyze double stars, moving about each other, according to the requirements of the gravitational law.

Figure 5–6. A double-star system.

SIX EASY PIECES

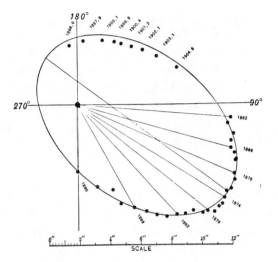

Figure 5–7. Orbit of Sirius *B* with respect to Sirius *A*.

That the law of gravitation is true at even bigger distances is indicated in Fig. 5–8. If one cannot see gravitation acting here, he has no soul. This figure shows one of the most beautiful things in the sky—a globular star cluster. All of the dots are stars. Although they look as if they are packed solid toward the center, that is due to the fallibility of our instruments. Actually, the distances between even the centermost stars are very great and they very rarely collide. There are more stars in the interior than farther out, and as we move outward there are fewer and fewer. It is obvious that there is an attraction among these stars. It is clear that gravitation exists at these enormous dimensions, perhaps 100,000 times the size of the solar system. Let us now go further, and look at an *entire galaxy*, shown in Fig. 5–9. The shape of this galaxy indicates an obvious tendency for its matter to agglomerate. Of course we cannot prove that the law here is precisely inverse square, only that there is still an attraction, at this enormous dimension, that holds the whole thing together. One may say, "Well, that is all very clever but why is it not just a ball?" Because it is *spinning* and has *angular momentum* which it cannot give up as it contracts; it must

The Theory of Gravitation

Figure 5–8. A globular star cluster.

contract mostly in a plane. (Incidentally, if you are looking for a good problem, the exact details of how the arms are formed and what determines the shapes of these galaxies has not been worked out.) It is, however, clear that the shape of the galaxy is due to gravitation even though the complexities of its structure have not yet allowed us to analyze it completely. In a galaxy we have a scale of perhaps 50,000 to 100,000 light years. The earth's distance from the sun is 8⅓ light *minutes*, so you can see how large these dimensions are.

Gravity appears to exist at even bigger dimensions, as indicated by Fig. 5–10, which shows many "little" things clustered together. This is a *cluster of galaxies*, just like a star cluster. Thus galaxies attract each other at such distances that they too are agglomerated into clusters. Perhaps gravitation exists even over distances of *tens of millions* of light years; so far as we now know, gravity seems to go out forever inversely as the square of the distance.

Not only can we understand the nebulae, but from the law of gravitation we can even get some ideas about the origin of the stars. If we have a big cloud of dust and gas, as indicated in Fig. 5–11, the gravitational attractions of the pieces of dust for one another might

Figure 5–9. A galaxy.

make them form little lumps. Barely visible in the figure are "little" black spots which may be the beginning of the accumulations of dust and gases which, due to their gravitation, begin to form stars. Whether we have ever seen a star form or not is still debatable. Figure 5–12 shows the one piece of evidence which suggests that we have. At the left is a picture of a region of gas with some stars in it taken in 1947, and at the right is another picture, taken only 7 years later, which shows two new bright spots. Has gas accumulated, has gravity acted hard enough and collected it into a ball big enough that the stellar nuclear reaction starts in the interior and turns it into a star? Perhaps, and perhaps not. It is unreasonable that in only seven years we should be so lucky as to see a star change itself into visible form; it is much less probable that we should see *two!*

Cavendish's experiment

Gravitation, therefore, extends over enormous distances. But if there is a force between *any* pair of objects, we ought to be able to measure the force between our own objects. Instead of having to

The Theory of Gravitation

Figure 5–10. A cluster of galaxies.

watch the stars go around each other, why can we not take a ball of lead and a marble and watch the marble go toward the ball of lead? The difficulty of this experiment when done in such a simple manner is the very weakness or delicacy of the force. It must be done with extreme care, which means covering the apparatus to keep the air out, making sure it is not electrically charged, and so on; then the force can be measured. It was first measured by Cavendish with an apparatus which is schematically indicated in Fig. 5–13. This first demonstrated the direct force between two large, fixed balls of lead and two smaller balls of lead on the ends of an arm supported by a very fine fiber, called a torsion fiber. By measuring how much the fiber gets twisted, one can measure the strength of the force, verify that it is inversely proportional to the square of the distance, and determine how strong it is. Thus, one may accurately determine the coefficient G in the formula

$$F = G\frac{mm'}{r^2}.$$

SIX EASY PIECES

Figure 5–11. An interstellar dust cloud.

All the masses and distances are known. You say, "We knew it already for the earth." Yes, but we did not know the *mass* of the earth. By knowing G from this experiment and by knowing how strongly the earth attracts, we can indirectly learn how great is the mass of the earth! This experiment has been called "weighing the earth." Cavendish claimed he was weighing the earth, but what he was measuring was the coefficient G of the gravity law. This is the only way in which the mass of the earth can be determined. G turns out to be

$$6.670 \times 10^{-11} \text{ newton} \cdot \text{m}^2/\text{kg}^2.$$

It is hard to exaggerate the importance of the effect on the history of science produced by this great success of the theory of gravitation. Compare the confusion, the lack of confidence, the incomplete knowledge that prevailed in the earlier ages, when there were endless debates and paradoxes, with the clarity and simplicity of this law—this fact that all the moons and planets and stars have

The Theory of Gravitation

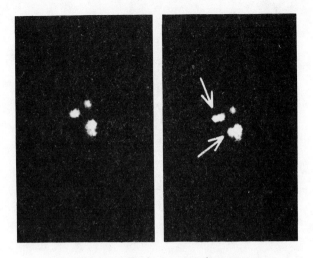

Figure 5–12. The formation of new stars?

such a *simple rule* to govern them, and further that man could *understand* it and deduce how the planets should move! This is the reason for the success of the sciences in following years, for it gave hope that the other phenomena of the world might also have such beautifully simple laws.

What is gravity?

But is this such a simple law? What about the machinery of it? All we have done is to describe *how* the earth moves around the sun, but we have not said *what makes it go*. Newton made no hypotheses about this; he was satisfied to find *what* it did without getting into the machinery of it. *No one has since given any machinery.* It is characteristic of the physical laws that they have this abstract character. The law of conservation of energy is a theorem concerning quantities that have to be calculated and added together, with no mention of the machinery, and likewise the great laws of mechanics are quantitative mathematical laws for which no machinery is available. Why can

SIX EASY PIECES

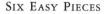
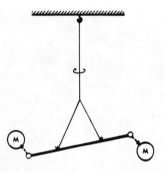

Figure 5–13. A simplified diagram of the apparatus used by Cavendish to verify the law of universal gravitation for small
objects and to measure the gravitational constant G.

we use mathematics to describe nature without a mechanism behind it? No one knows. We have to keep going because we find out
more that way.

Many mechanisms for gravitation have been suggested. It is
interesting to consider one of these, which many people have
thought of from time to time. At first, one is quite excited and
happy when he "discovers" it, but he soon finds that it is not correct.
It was first discovered about 1750. Suppose there were many particles moving in space at a very high speed in all directions and being
only slightly absorbed in going through matter. When they *are*
absorbed, they give an impulse to the earth. However, since there
are as many going one way as another, the impulses all balance. But
when the sun is nearby, the particles coming toward the earth
through the sun are partially absorbed, so fewer of them are coming
from the sun than are coming from the other side. Therefore, the
earth feels a net impulse toward the sun and it does not take one
long to see that it is inversely as the square of the distance—because
of the variation of the solid angle that the sun subtends as we vary
the distance. What is wrong with that machinery? It involves some
new consequences which are *not true*. This particular idea has the

The Theory of Gravitation

following trouble: the earth, in moving around the sun, would impinge on more particles which are coming from its forward side than from its hind side (when you run in the rain, the rain in your face is stronger than that on the back of your head!). Therefore there would be more impulse given the earth from the front, and the earth would feel a *resistance to motion* and would be slowing up in its orbit. One can calculate how long it would take for the earth to stop as a result of this resistance, and it would not take long enough for the earth to still be in its orbit, so this mechanism does not work. No machinery has ever been invented that "explains" gravity without also predicting some other phenomenon that does *not* exist.

Next we shall discuss the possible relation of gravitation to other forces. There is no explanation of gravitation in terms of other forces at the present time. It is not an aspect of electricity or anything like that, so we have no explanation. However, gravitation and other forces are very similar, and it is interesting to note analogies. For example, the force of electricity between two charged objects looks just like the law of gravitation: the force of electricity is a constant, with a minus sign, times the product of the charges, and varies inversely as the square of the distance. It is in the opposite direction—likes repel. But is it still not very remarkable that the two laws involve the same function of distance? Perhaps gravitation and electricity are much more closely related than we think. Many attempts have been made to unify them; the so-called unified field theory is only a very elegant attempt to combine electricity and gravitation; but, in comparing gravitation and electricity, the most interesting thing is the *relative strengths* of the forces. Any theory that contains them both must also deduce how strong the gravity is.

If we take, in some natural units, the repulsion of two electrons (nature's universal charge) due to electricity, and the attraction of two electrons due to their masses, we can measure the ratio of electrical repulsion to the gravitational attraction. The ratio is independent of the distance and is a fundamental constant of nature.

The Theory of Gravitation

universe got older the ratio of the age of the universe to the time which it takes for light to go across a proton would be gradually increasing. Is it possible that the gravitational constant *is* changing with time? Of course the changes would be so small that it is quite difficult to be sure.

One test which we can think of is to determine what would have been the effect of the change during the past 10^9 years, which is approximately the age from the earliest life on the earth to now, and one-tenth of the age of the universe. In this time, the gravity constant would have increased by about 10 percent. It turns out that if we consider the structure of the sun—the balance between the weight of its material and the rate at which radiant energy is generated inside it—we can deduce that if the gravity were 10 percent stronger, the sun would be much more than 10 percent brighter—by the *sixth power* of the gravity constant! If we calculate what happens to the orbit of the earth when the gravity is changing, we find that the earth was then *closer in*. Altogether, the earth would be about 100 degrees centigrade hotter, and all of the water would not have been in the sea, but vapor in the air, so life would not have started in the sea. So we do *not* now believe that the gravity constant is changing with the age of the universe. But such arguments as the one we have just given are not very convincing, and the subject is not completely closed.

It is a fact that the force of gravitation is proportional to the *mass*, the quantity which is fundamentally a measure of *inertia*—of how hard it is to hold something which is going around in a circle. Therefore two objects, one heavy and one light, going around a larger object in the same circle at the same speed because of gravity, will stay together because to go in a circle *requires* a force which is stronger for a bigger mass. That is, the gravity is stronger for a given mass in *just the right proportion* so that the two objects will go around together. If one object were inside the other it would *stay* inside; it is a perfect balance. Therefore, Gagarin or Titov would find things "weightless" inside a space ship; if they happened to let go of a piece of chalk, for example, it would go around the earth in exactly the same way as the whole space ship, and so it would appear to remain suspended before them in space. It is very interesting that this force is *exactly* proportional to the mass with great precision, because if it were not exactly proportional there would be some effect by which inertia and weight would differ. The absence of such an effect has been checked with great accuracy by an experiment done first by Eötvös in 1909 and more recently by Dicke. For all substances tried, the masses and weights are exactly proportional within 1 part in 1,000,000,000, or less. This is a remarkable experiment.

Gravity and relativity

Another topic deserving discussion is Einstein's modification of Newton's law of gravitation. In spite of all the excitement it created, Newton's law of gravitation is not correct! It was modified by Einstein to take into account the theory of relativity. According to Newton, the gravitational effect is instantaneous, that is, if we were to move a mass, we would at once feel a new force because of the new position of that mass; by such means we could send signals at infinite speed. Einstein advanced arguments which suggest that we *cannot send signals faster than the speed of light*, so the law of gravitation must be wrong. By correcting it to take the delays into account, we have a new law, called Einstein's law of gravitation. One feature of this new law which is quite easy to understand is this: In the Einstein relativity theory, anything which has *energy* has mass—mass in the sense that it is attracted gravitationally. Even light, which has an energy, has a "mass." When a light beam, which has energy in it, comes past the sun there is an attraction on it by the sun. Thus the light does not go straight, but is deflected. During the eclipse of the sun, for example, the stars which are around the sun should appear displaced from where they would be if the sun were not there, and this has been observed.

Finally, let us compare gravitation with other theories. In recent years we have discovered that all mass is made of tiny particles and that there are several kinds of interactions, such as nuclear forces, etc. None of these nuclear or electrical forces has yet been found to explain gravitation. The quantum-mechanical aspects of nature have not yet been carried over to gravitation. When the scale is so small that we need the quantum effects, the gravitational effects are so weak that the need for a quantum theory of gravitation has not yet developed. On the other hand, for consistency in our physical theories it would be important to see whether Newton's law modified to Einstein's law can be further modified to be consistent with the uncertainty principle. This last modification has not yet been completed.

Six

QUANTUM BEHAVIOR

Atomic mechanics

◐ In the last few chapters we have treated the essential ideas necessary for an understanding of most of the important phenomena of light—or electromagnetic radiation in general. (We have left a few special topics for next year. Specifically, the theory of the index of dense materials and total internal reflection.) What we have dealt with is called the "classical theory" of electric waves, which turns out to be a completely adequate description of nature for a large number of effects. We have not had to worry yet about the fact that light energy comes in lumps or "photons."

We would like to take up as our next subject the problem of the behavior of relatively large pieces of matter—their mechanical and thermal properties, for instance. In discussing these, we will find that the "classical" (or older) theory fails almost immediately, because matter is really made up of atomic-sized particles. Still, we will deal only with the classical part, because that is the only part that we can understand using the classical mechanics we have been learning. But we shall not be very successful. We shall find that in the case of matter, unlike the case of light, we shall be in difficulty relatively soon. We could, of course, continuously skirt away from the atomic effects, but we shall instead interpose here a short excursion in which we will describe the basic ideas of the quantum properties of matter, i.e., the quantum ideas of atomic physics, so that you will have some feeling for what it is we are leaving out. For we will have to leave out some important subjects that we cannot avoid coming close to.

So we will give now the *introduction* to the subject of quantum mechanics, but will not be able actually to get into the subject until much later.

"Quantum mechanics" is the description of the behavior of matter in all its details and, in particular, of the happenings on an atomic scale. Things on a very small scale behave like nothing that you have any direct experience about. They do not behave like waves, they do not behave like particles, they do not behave like clouds, or billiard balls, or weights on springs, or like anything that you have ever seen.

Newton thought that light was made up of particles, but then it was discovered, as we have seen here, that it behaves like a wave. Later, however (in the beginning of the twentieth century) it was found that light did indeed sometimes behave like a particle. Historically, the electron, for example, was thought to behave like a particle, and then it was found that in many respects it behaved like a wave. So it really behaves like neither. Now we have given up. We say: "It is like *neither*."

There is one lucky break, however—electrons behave just like light. The quantum behavior of atomic objects (electrons, protons, neutrons, photons, and so on) is the same for all; they are all "particle waves," or whatever you want to call them. So what we learn about the properties of electrons (which we shall use for our examples) will apply also to all "particles," including photons of light.

The gradual accumulation of information about atomic and small-scale behavior during the first quarter of this century, which gave some indications about how small things do behave, produced an increasing confusion which was finally resolved in 1926 and 1927 by Schrödinger, Heisenberg, and Born. They finally obtained a consistent description of the behavior of matter on a small scale. We take up the main features of that description in this chapter.

Quantum Behavior

Because atomic behavior is so unlike ordinary experience, it is very difficult to get used to and it appears peculiar and mysterious to everyone, both to the novice and to the experienced physicist. Even the experts do not understand it the way they would like to, and it is perfectly reasonable that they should not, because all of direct, human experience and of human intuition applies to large objects. We know how large objects will act, but things on a small scale just do not act that way. So we have to learn about them in a sort of abstract or imaginative fashion and not by connection with our direct experience.

In this chapter we shall tackle immediately the basic element of the mysterious behavior in its most strange form. We choose to examine a phenomenon which is impossible, *absolutely* impossible, to explain in any classical way, and which has in it the heart of quantum mechanics. In reality, it contains the *only* mystery. We cannot explain the mystery in the sense of "explaining" how it works. We will *tell* you how it works. In telling you how it works we will have told you about the basic peculiarities of all quantum mechanics.

An experiment with bullets

To try to understand the quantum behavior of electrons, we shall compare and contrast their behavior, in a particular experimental setup, with the more familiar behavior of particles like bullets, and with the behavior of waves like water waves. We consider first the behavior of bullets in the experimental setup shown diagrammatically in Fig. 6–1. We have a machine gun that shoots a stream of bullets. It is not a very good gun, in that it sprays the bullets (randomly) over a fairly large angular spread, as indicated in the figure. In front of the gun we have a wall (made of armor plate) that has in it two holes just about big enough to let a bullet through. Beyond the wall is a backstop (say a thick wall of wood) which will "absorb" the bullets when they hit it. In front of the wall we have an object which we shall call a "detector" of bullets. It might be a box

SIX EASY PIECES

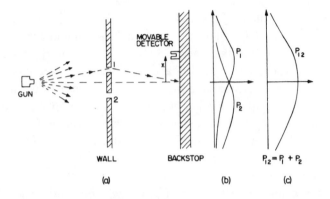

Figure 6-1. Interference experiment with bullets.

containing sand. Any bullet that enters the detector will be stopped and accumulated. When we wish, we can empty the box and count the number of bullets that have been caught. The detector can be moved back and forth (in what we will call the x-direction). With this apparatus, we can find out experimentally the answer to the question: "What is the probability that a bullet which passes through the holes in the wall will arrive at the backstop at the distance x from the center?" First, you should realize that we should talk about probability, because we cannot say definitely where any particular bullet will go. A bullet which happens to hit one of the holes may bounce off the edges of the hole, and may end up anywhere at all. By "probability" we mean the chance that the bullet will arrive at the detector, which we can measure by counting the number which arrive at the detector in a certain time and then taking the ratio of this number to the *total* number that hit the backstop during that time. Or, if we assume that the gun always shoots at the same rate during the measurements, the probability we want is just proportional to the number that reach the detector in some standard time interval.

For our present purposes we would like to imagine a somewhat idealized experiment in which the bullets are not real bullets, but

Quantum Behavior

are *indestructible* bullets—they cannot break in half. In our experiment we find that bullets always arrive in lumps, and when we find something in the detector, it is always one whole bullet. If the rate at which the machine gun fires is made very low, we find that at any given moment either nothing arrives, or one and only one—exactly one—bullet arrives at the backstop. Also, the size of the lump certainly does not depend on the rate of firing of the gun. We shall say: "Bullets *always* arrive in identical lumps." What we measure with our detector is the probability of arrival of a lump. And we measure the probability as a function of x. The result of such measurements with this apparatus (we have not yet done the experiment, so we are really imagining the result) is plotted in the graph drawn in part (c) of Fig. 6–1. In the graph we plot the probability to the right and x vertically, so that the x-scale fits the diagram of the apparatus. We call the probability P_{12} because the bullets may have come either through hole 1 or through hole 2. You will not be surprised that P_{12} is large near the middle of the graph but gets small if x is very large. You may wonder, however, why P_{12} has its maximum value at $x = 0$. We can understand this fact if we do our experiment again after covering up hole 2, and once more while covering up hole 1. When hole 2 is covered, bullets can pass only through hole 1, and we get the curve marked P_1 in part (b) of the figure. As you would expect, the maximum of P_1 occurs at the value of x which is on a straight line with the gun and hole 1. When hole 1 is closed, we get the symmetric curve P_2 drawn in the figure. P_2 is the probability distribution for bullets that pass through hole 2. Comparing parts (b) and (c) of Fig. 6–1, we find the important result that

$$P_{12} = P_1 + P_2. \tag{6.1}$$

The probabilities just add together. The effect with both holes open is the sum of the effects with each hole open alone. We shall call this result an observation of "*no interference*," for a reason that you will see later. So much for bullets. They come in lumps, and their probability of arrival shows no interference.

An experiment with waves

Now we wish to consider an experiment with water waves. The apparatus is shown diagrammatically in Fig. 6–2. We have a shallow trough of water. A small object labeled the "wave source" is jiggled up and down by a motor and makes circular waves. To the right of the source we have again a wall with two holes, and beyond that is a second wall, which, to keep things simple, is an "absorber," so that there is no reflection of the waves that arrive there. This can be done by building a gradual sand "beach." In front of the beach we place a detector which can be moved back and forth in the x-direction, as before. The detector is now a device which measures the "intensity" of the wave motion. You can imagine a gadget which measures the height of the wave motion, but whose scale is calibrated in proportion to the *square* of the actual height, so that the reading is proportional to the intensity of the wave. Our detector reads, then, in proportion to the *energy* being carried by the wave—or rather, the rate at which energy is carried to the detector.

With our wave apparatus, the first thing to notice is that the intensity can have *any* size. If the source just moves a very small

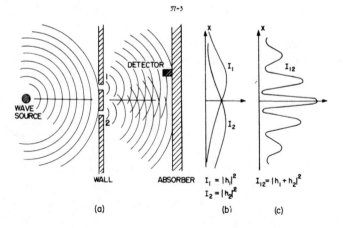

Figure 6–2. Interference experiment with water waves.

Quantum Behavior

amount, then there is just a little bit of wave motion at the detector. When there is more motion at the source, there is more intensity at the detector. The intensity of the wave can have any value at all. We would *not* say that there was any "lumpiness" in the wave intensity.

Now let us measure the wave intensity for various values of x (keeping the wave source operating always in the same way). We get the interesting-looking curve marked I_{12} in part (c) of the figure.

We have already worked out how such patterns can come about when we studied the interference of electric waves. In this case we would observe that the original wave is diffracted at the holes, and new circular waves spread out from each hole. If we cover one hole at a time and measure the intensity distribution at the absorber we find the rather simple intensity curves shown in part (b) of the figure. I_1 is the intensity of the wave from hole 1 (which we find by measuring when hole 2 is blocked off) and I_2 is the intensity of the wave from hole 2 (seen when hole 1 is blocked).

The intensity I_{12} observed when both holes are open is certainly *not* the sum of I_1 and I_2. We say that there is "interference" of the two waves. At some places (where the curve I_{12} has its maxima) the waves are "in phase" and the wave peaks add together to give a large amplitude and, therefore, a large intensity. We say that the two waves are "interfering constructively" at such places. There will be such constructive interference wherever the distance from the detector to one hole is a whole number of wavelengths larger (or shorter) than the distance from the detector to the other hole.

At those places where the two waves arrive at the detector with a phase difference of π (where they are "out of phase") the resulting wave motion at the detector will be the difference of the two amplitudes. The waves "interfere destructively," and we get a low value for the wave intensity. We expect such low values wherever the distance between hole 1 and the detector is different from the distance between hole 2 and the detector by an odd number of half-wavelengths. The low values of I_{12} in Fig. 6–2 correspond to the places where the two waves interfere destructively.

You will remember that the quantitative relationship between I_1,

I_2, and I_{12} can be expressed in the following way: The instantaneous height of the water wave at the detector for the wave from hole 1 can be written as (the real part of) $\hat{h}_1 e^{i\omega t}$, where the "amplitude" $\hat{h}_1$ is, in general, a complex number. The intensity is proportional to the mean squared height or, when we use the complex numbers, to $|\hat{h}_1|^2$. Similarly, for hole 2 the height is $\hat{h}_2 e^{i\omega t}$ and the intensity is proportional to $|\hat{h}_2|^2$. When both holes are open, the wave heights add to give the height $(\hat{h}_1 + \hat{h}_2)e^{i\omega t}$ and the intensity $|\hat{h}_1 + \hat{h}_2|^2$. Omitting the constant of proportionality for our present purposes, the proper relations for *interfering waves* are

$$I_1 = |\hat{h}_1|^2, \qquad I_2 = |\hat{h}_2|^2, \qquad I_{12} = |\hat{h}_1 + \hat{h}_2|^2. \qquad (6.2)$$

You will notice that the result is quite different from that obtained with bullets (Eq. 6.1). If we expand $|\hat{h}_1 + \hat{h}_2|^2$ we see that

$$|\hat{h}_1 + \hat{h}_2|^2 = |\hat{h}_1|^2 + |\hat{h}_2|^2 + 2|\hat{h}_1||\hat{h}_2| \cos \delta, \qquad (6.3)$$

where δ is the phase difference between $\hat{h}_1$ and $\hat{h}_2$. In terms of the intensities, we could write

$$I_{12} = I_1 + I_2 + 2\sqrt{I_1 I_2} \cos \delta. \qquad (6.4)$$

The last term in (6.4) is the "interference term." So much for water waves. The intensity can have any value, and it shows inteference.

An experiment with electrons

Now we imagine a similar experiment with electrons. It is shown diagrammatically in Fig. 6–3. We make an electron gun which consists of a tungsten wire heated by an electric current and surrounded by a metal box with a hole in it. If the wire is at a negative voltage with respect to the box, electrons emitted by the wire will be accelerated toward the walls and some will pass through the hole. All the electrons which come out of the gun will have (nearly) the same energy.

Quantum Behavior

In front of the gun is again a wall (just a thin metal plate) with two holes in it. Beyond the wall is another plate which will serve as a "backstop." In front of the backstop we place a movable detector. The detector might be a geiger counter or, perhaps better, an electron multiplier, which is connected to a loudspeaker.

We should say right away that you should not try to set up this experiment (as you could have done with the two we have already described). This experiment has never been done in just this way. The trouble is that the apparatus would have to be made on an impossibly small scale to show the effects we are interested in. We are doing a "thought experiment," which we have chosen because it is easy to think about. We know the results that *would* be obtained because there *are* many experiments that have been done, in which the scale and the proportions have been chosen to show the effects we shall describe.

The first thing we notice with our electron experiment is that we hear sharp "clicks" from the detector (that is, from the loudspeaker). And all "clicks" are the same. There are *no* "half-clicks."

We would also notice that the "clicks" come very erratically. Something like: click . . . . . click-click . . . click . . . . click . . . .

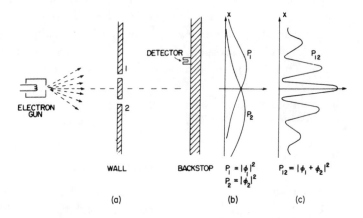

Figure 6–3. Interference experiment with electrons.

click-click . . . . . . click . . . , etc., just as you have, no doubt, heard a geiger counter operating. If we count the clicks which arrive in a sufficiently long time—say for many minutes—and then count again for another equal period, we find that the two numbers are very nearly the same. So we can speak of the *average rate* at which the clicks are heard (so-and-so-many clicks per minute on the average).

As we move the detector around, the *rate* at which the clicks appear is faster or slower, but the size (loudness) of each click is always the same. If we lower the temperature of the wire in the gun the rate of clicking slows down, but still each click sounds the same. We would notice also that if we put two separate detectors at the backstop, one *or* the other would click, but never both at once. (Except that once in a while, if there were two clicks very close together in time, our ear might not sense the separation.) We conclude, therefore, that whatever arrives at the backstop arrives in "lumps." All the "lumps" are the same size: only whole "lumps" arrive, and they arrive one at a time at the backstop. We shall say: "Electrons always arrive in identical lumps."

Just as for our experiment with bullets, we can now proceed to find experimentally the answer to the question: "What is the relative probability that an electron 'lump' will arrive at the backstop at various distances x from the center?" As before, we obtain the relative probability by observing the rate of clicks, holding the operation of the gun constant. The probability that lumps will arrive at a particular x is proportional to the average rate of clicks at that x.

The result of our experiment is the interesting curve marked P_{12} in part (c) of Fig. 6–3. Yes! That is the way electrons go.

The interference of electron waves

Now let us try to analyze the curve of Fig. 6–3 to see whether we can understand the behavior of the electrons. The first thing we would say is that since they come in lumps, each lump, which we may as

Quantum Behavior

well call an electron, has come either through hole 1 or through hole 2. Let us write this in the form of a "Proposition":

Proposition A: Each electron *either* goes through hole 1 *or* it goes through hole 2.

Assuming Proposition A, all electrons that arrive at the backstop can be divided into two classes: (1) those that come through hole 1, and (2) those that come through hole 2. So our observed curve must be the sum of the effects of the electrons which come through hole 1 and the electrons which come through hole 2. Let us check this idea by experiment. First, we will make a measurement for those electrons that come through hole 1. We block off hole 2 and make our counts of the clicks from the detector. From the clicking rate, we get P_1. The result of the measurement is shown by the curve marked P_1 in part (b) of Fig. 6–3. The result seems quite reasonable. In a similar way, we measure P_2, the probability distribution for the electrons that come through hole 2. The result of this measurement is also drawn in the figure.

The result P_{12} obtained with *both* holes open is clearly not the sum of P_1 and P_2, the probabilities for each hole alone. In analogy with our water-wave experiment, we say: "There is interference."

$$\textit{For electrons:} \qquad P_{12} \neq P_1 + P_2. \qquad (6.5)$$

How can such an interference come about? Perhaps we should say: "Well, that means, presumably, that it is *not true* that the lumps go either through hole 1 or hole 2, because if they did, the probabilities should add. Perhaps they go in a more complicated way. They split in half and . . ." But no! They cannot, they always arrive in lumps . . . "Well, perhaps some of them go through 1, and then they go around through 2, and then around a few more times, or by some other complicated path . . . then by closing hole 2, we changed the chance that an electron that *started out* through hole 1 would finally get to the backstop . . ." But notice! There are some points at which very few electrons arrive when *both* holes are open, but which receive many electrons if we close one hole, so *closing* one hole *increased* the number from the other. Notice, however, that at the center of the pattern, P_{12} is more than twice as large as $P_1 + P_2$. It is as though closing one hole *decreased* the number of electrons which come through the other hole. It seems hard to explain *both* effects by proposing that the electrons travel in complicated paths.

It is all quite mysterious. And the more you look at it the more mysterious it seems. Many ideas have been concocted to try to explain the curve for P_{12} in terms of individual electrons going around in complicated ways through the holes. None of them has succeeded. None of them can get the right curve for P_{12} in terms of P_1 and P_2.

Yet, surprisingly enough, the *mathematics* for relating P_1 and P_2 to P_{12} is extremely simple. For P_{12} is just like the curve I_{12} of Fig. 6–2, and *that* was simple. What is going on at the backstop can be described by two complex numbers that we can call $\hat{\phi}_1$ and $\hat{\phi}_2$ (they are functions of x, of course). The absolute square of $\hat{\phi}_1$ gives the effect with only hole 1 open. That is, $P_1 = |\hat{\phi}_1|^2$. The effect with only hole 2 open is given by $\hat{\phi}_2$ in the same way. That is, $P_2 = |\hat{\phi}_1|^2$. And the combined effect of the two holes is just $P_{12} = |\hat{\phi}_1 + \hat{\phi}_2|^2$. The *mathematics* is the same as what we had for the water waves! (It is hard to see how one could get such a simple result from a complicated game of electrons going back and forth through the plate on some strange trajectory.)

We conclude the following: The electrons arrive in lumps, like particles, and the probability of arrival of these lumps is distributed like the distribution of intensity of a wave. It is in this sense that an electron behaves "sometimes like a particle and sometimes like a wave."

Incidentally, when we were dealing with classical waves we defined the intensity as the mean over time of the square of the wave amplitude, and we used complex numbers as a mathematical trick to simplify the analysis. But in quantum mechanics it turns out that the amplitudes *must* be represented by complex numbers. The real parts alone will not do. That is a technical point, for the moment, because the formulas look just the same.

Since the probability of arrival through both holes is given so simply, although it is not equal to $(P_1 + P_2)$, that is really all there is to say. But there are a large number of subtleties involved in the fact that nature does work this way. We would like to illustrate some of these subtleties for you now. First, since the number that arrives at a particular point is *not* equal to the number that arrives through 1 plus the number that arrives through 2, as we would have concluded from Proposition A, undoubtedly we should conclude that *Proposition A is false*. It is *not* true that the electrons go *either* through hole 1 or hole 2. But that conclusion can be tested by another experiment.

Watching the electrons

We shall now try the following experiment. To our electron apparatus we add a very strong light source, placed behind the wall and between the two holes, as shown in Fig. 6–4. We know that electric charges scatter light. So when an electron passes, however it does pass, on its way to the detector, it will scatter some light to our eye, and we can *see* where the electron goes. If, for instance, an electron were to take the path via hole 2 that is sketched in Fig. 6–4, we should see a flash of light coming from the vicinity of the place marked A in the figure. If an electron passes through hole 1 we would expect to see a flash from the vicinity of the upper hole. If it should happen that we get light from both places at the same time, because the electron divides in half . . . Let us just do the experiment!

Here is what we see: *every* time that we hear a "click" from our electron detector (at the backstop), we *also see* a flash of light *either* near hole 1 *or* near hole 2, but *never* both at once! And we observe the same result no matter where we put the detector. From this observation we conclude that when we look at the electrons we find that the electrons go either through one hole or the other. Experimentally, Proposition A is necessarily true.

SIX EASY PIECES

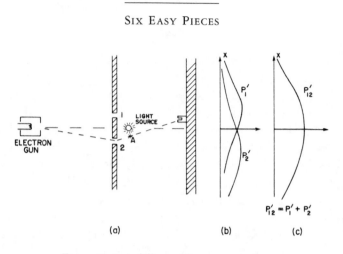

Figure 6–4. A different electron experiment.

What, then, is wrong with our argument *against* Proposition A? Why *isn't* P_{12} just equal to $P_1 + P_2$? Back to experiment! Let us keep track of the electrons and find out what they are doing. For each position (x-location) of the detector we will count the electrons that arrive and *also* keep track of which hole they went through, by watching for the flashes. We can keep track of things this way: whenever we hear a "click" we will put a count in Column 1 if we see the flash near hole 1, and if we see the flash near hole 2, we will record a count in Column 2. Every electron which arrives is recorded in one of two classes: those which come through 1 and those which come through 2. From the number recorded in Column 1 we get the probability P'_1 that an electron will arrive at the detector via hole 1; and from the number recorded in Column 2 we get P'_2, the probability that an electron will arrive at the detector via hole 2. If we now repeat such a measurement for many values of x, we get the curves for P'_1 and P'_2 shown in part (b) of Fig. 6–4.

Well, that is not too surprising! We get for P'_1 something quite similar to what we got before for P_1 by blocking off hole 2; and P'_2 is similar to what we got by blocking hole 1. So there is *not* any complicated business like going through both holes. When we watch them, the electrons come through just as we would expect

Quantum Behavior

them to come through. Whether the holes are closed or open, those which we see come through hole 1 are distributed in the same way whether hole 2 is open or closed.

But wait! What do we have *now* for the *total* probability, the probability that an electron will arrive at the detector by any route? We already have that information. We just pretend that we never looked at the light flashes, and we lump together the detector clicks which we have separated into the two columns. We *must* just *add* the numbers. For the probability that an electron will arrive at the backstop by passing through *either* hole, we do find $P'_{12} = P_1 + P_2$. That is, although we succeeded in watching which hole our electrons come through, we no longer get the old interference curve P_{12}, but a new one, P'_{12}, showing no interference! If we turn out the light P_{12} is restored.

We must conclude that *when we look at the electrons* the distribution of them on the screen is different than when we do not look. Perhaps it is turning on our light source that disturbs things? It must be that the electrons are very delicate, and the light, when it scatters off the electrons, gives them a jolt that changes their motion. We know that the electric field of the light acting on a charge will exert a force on it. So perhaps we *should* expect the motion to be changed. Anyway, the light exerts a big influence on the electrons. By trying to "watch" the electrons we have changed their motions. That is, the jolt given to the electron when the photon is scattered by it is such as to change the electron's motion enough so that if it *might* have gone to where P_{12} was at a maximum it will instead land where P_{12} was a minimum; that is why we no longer see the wavy interference effects.

You may be thinking: "Don't use such a bright source! Turn the brightness down! The light waves will then be weaker and will not disturb the electrons so much. Surely, by making the light dimmer and dimmer, eventually the wave will be weak enough that it will have a negligible effect." O.K. Let's try it. The first thing we observe is that the flashes of light scattered from the electrons as they pass by does *not* get weaker. *It is always the same-sized flash.* The only thing that happens as the light is made dimmer is that sometimes we hear a "click" from the detector but see *no flash at all*. The electron has gone by without being "seen." What we are observing is that light *also* acts like electrons; we *knew* that it was "wavy," but now we find that it is also "lumpy." It always arrives—or is scattered—in lumps that we call "photons." As we turn down the *intensity* of the light source we do not change the *size* of the photons, only the *rate* at which they are emitted. *That* explains why, when our source is dim, some electrons get by without being seen. There did not happen to be a photon around at the time the electron went through.

This is all a little discouraging. If it is true that whenever we "see" the electron we see the same-sized flash, then those electrons we see are *always* the disturbed ones. Let us try the experiment with a dim light anyway. Now whenever we hear a click in the detector we will keep a count in three columns: in Column (1) those electrons seen by hole 1, in Column (2) those electrons seen by hole 2, and in Column (3) those electrons not seen at all. When we work up our data (computing the probabilities) we find these results: Those "seen by hole 1" have a distribution like P'_1; those "seen by hole 2" have a distribution like P'_2 (so that those "seen by either hole 1 or 2" have a distribution like P'_{12}); and those "not seen at all" have a "wavy" distribution just like P_{12} of Fig. 6–3! *If the electrons are not seen, we have interference!*

That is understandable. When we do not see the electron, no photon disturbs it, and when we do see it, a photon has disturbed it. There is always the same amount of disturbance because the light photons all produce the same-sized effects and the effect of the photons being scattered is enough to smear out any interference effect.

Is there not *some* way we can see the electrons without disturbing them? We learned in an earlier chapter that the momentum carried by a "photon" is inversely proportional to its wavelength ($p = h/\lambda$). Certainly the jolt given to the electron when the photon is scattered toward our eye depends on the momentum that photon carries.

Quantum Behavior

Aha! If we want to disturb the electrons only slightly we should not have lowered the *intensity* of the light, we should have lowered its *frequency* (the same as increasing its wavelength). Let us use light of a redder color. We could even use infrared light, or radiowaves (like radar), and "see" where the electron went with the help of some equipment that can "see" light of these longer wavelengths. If we use "gentler" light perhaps we can avoid disturbing the electrons so much.

Let us try the experiment with longer waves. We shall keep repeating our experiment, each time with light of a longer wavelength. At first, nothing seems to change. The results are the same. Then a terrible thing happens. You remember that when we discussed the microscope we pointed out that, due to the *wave nature* of the light, there is a limitation on how close two spots can be and still be seen as two separate spots. This distance is of the order of the wavelength of light. So now, when we make the wavelength longer than the distance between our holes, we see a *big* fuzzy flash when the light is scattered by the electrons. We can no longer tell which hole the electron went through! We just know it went somewhere! And it is just with light of this color that we find that the jolts given to the electron are small enough so that P'_{12} begins to look like P_{12}—that we begin to get some interference effect. And it is only for wavelengths much longer than the separation of the two holes (when we have no chance at all of telling where the electron went) that the disturbance due to the light gets sufficiently small that we again get the curve P_{12} shown in Fig. 6-3.

In our experiment we find that it is impossible to arrange the light in such a way that one can tell which hole the electron went through, and at the same time not disturb the pattern. It was suggested by Heisenberg that the then new laws of nature could only be consistent if there were some basic limitation on our experimental capabilities not previously recognized. He proposed, as a general principle, his *uncertainty principle*, which we can state in terms of our experiment as follows: "It is impossible to design an apparatus to determine which hole the electron passes through, that will not at the same time disturb the electrons enough to destroy the interference pattern." If an apparatus is capable of determining which hole the electron goes through, it *cannot* be so delicate that it does not disturb the pattern in an essential way. No one has ever found (or even thought of) a way around the uncertainty principle. So we must assume that it describes a basic characteristic of nature.

The complete theory of quantum mechanics which we now use to describe atoms and, in fact, all matter depends on the correctness of the uncertainty principle. Since quantum mechanics is such a successful theory, our belief in the uncertainty principle is reinforced. But if a way to "beat" the uncertainty principle were ever discovered, quantum mechanics would give inconsistent results and would have to be discarded as a valid theory of nature.

"Well," you say, "what about Proposition A? It is true, or is it *not* true, that the electron either goes through hole 1 or it goes through hole 2?" The only answer that can be given is that we have found from experiment that there is a certain special way that we have to think in order that we do not get into inconsistencies. What we must say (to avoid making wrong predictions) is the following: If one looks at the holes or, more accurately, if one has a piece of apparatus which is capable of determining whether the electrons go through hole 1 or hole 2, then one *can* say that it goes either through hole 1 or hole 2. *But*, when one does *not* try to tell which way the electron goes, when there is nothing in the experiment to disturb the electrons, then one may *not* say that an electron goes either through hole 1 or hole 2. If one does say that, and starts to make any deductions from the statement, he will make errors in the analysis. This is the logical tightrope on which we must walk if we wish to describe nature successfully.

◊   ◊   ◊

If the motion of all matter—as well as electrons—must be described in terms of waves, what about the bullets in our first experiment? Why didn't we see an interference pattern there? It turns out that for the bullets the wavelengths were so tiny that the

Quantum Behavior

interference patterns became very fine. So fine, in fact, that with any detector of finite size one could not distinguish the separate maxima and minima. What we saw was only a kind of average, which is the classical curve. In Fig. 6–5 we have tried to indicate schematically what happens with large-scale objects. Part (a) of the figure shows the probability distribution one might predict for bullets, using quantum mechanics. The rapid wiggles are supposed to represent the interference pattern one gets for waves of very short wavelength. Any physical detector, however, straddles several wiggles of the probability curve, so that the measurements show the smooth curve drawn in part (b) of the figure.

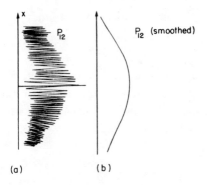

Figure 6–5. Interference pattern with bullets: (a) actual (schematic), (b) observed.

First principles of quantum mechanics

We will now write a summary of the main conclusions of our experiments. We will, however, put the results in a form which makes them true for a general class of such experiments. We can write our summary more simply if we first define an "ideal experiment" as one in which there are no uncertain external influences, i.e., no jiggling or other things going on that we cannot take into account. We would be quite precise if we said: "An ideal experiment is one in which all of the initial and final conditions of the experiment are completely specified." What we will call "an event" is, in general, just a specific set of initial and final conditions. (For example: "an electron leaves the gun, arrives at the detector, and nothing else happens.") Now for our summary.

SUMMARY

(1) The probability of an event in an ideal experiment is given by the square of the absolute value of a complex number ø which is called the probability amplitude.

$$P = \text{probability,}$$
$$\phi = \text{probability amplitude,} \qquad (6.6)$$
$$P = |\phi|^2.$$

(2) When an event can occur in several alternative ways, the probability amplitude for the event is the sum of the probability amplitudes for each way considered separately. There is interference.

$$\phi = \phi_1 + \phi_2,$$
$$P = |\phi_1 + \phi_2|^2. \qquad (6.7)$$

(3) If an experiment is performed which is capable of determining whether one or another alternative is actually taken, the probability of the event is the sum of the probabilities for each alternative. The interference is lost.

$$P = P_1 + P_2. \qquad (6.8)$$

One might still like to ask: "How does it work? What is the machinery behind the law?" No one has found any machinery behind the law. No one can "explain" any more than we have just

Quantum Behavior

"explained." No one will give you any deeper representation of the situation. We have no ideas about a more basic mechanism from which these results can be deduced.

We would like to emphasize a very important difference between classical and quantum mechanics. We have been talking about the probability that an electron will arrive in a given circumstance. We have implied that in our experimental arrangement (or even in the best possible one) it would be impossible to predict exactly what would happen. We can only predict the odds! This would mean, if it were true, that physics has given up on the problem of trying to predict exactly what will happen in a definite circumstance. Yes! Physics *has* given up. *We do not know how to predict what would happen in a given circumstance,* and we believe now that it is impossible, that the only thing that can be predicted is the probability of different events. It must be recognized that this is a retrenchment in our earlier ideal of understanding nature. It may be a backward step, but no one has seen a way to avoid it.

We make now a few remarks on a suggestion that has sometimes been made to try to avoid the description we have given: "Perhaps the electron has some kind of internal works—some inner variables—that we do not yet know about. Perhaps that is why we cannot predict what will happen. If we could look more closely at the electron we could be able to tell where it would end up." So far as we know, that is impossible. We would still be in difficulty. Suppose we were to assume that inside the electron there is some kind of machinery that determines where it is going to end up. That machine must *also* determine which hole it is going to go through on its way. But we must not forget that what is inside the electron should not be dependent on what *we* do, and in particular upon whether we open or close one of the holes. So if an electron, before it starts, has already made up its mind (a) which hole it is going to use, and (b) where it is going to land, we should find P_1 for those electrons that have chosen hole 1, P_2 for those that have chosen hole 2, *and necessarily* the sum $P_1 + P_2$ for those that arrive through the two holes. There seems to be no way around this. But we have verified experimentally that that is not the case. And no one has figured a way out of this puzzle. So at the present time we must limit ourselves to computing probabilities. We say "at the present time," but we suspect very strongly that it is something that will be with us forever—that it is impossible to beat that puzzle—that this is the way nature really *is*.

The uncertainty principle

This is the way Heisenberg stated the uncertainty principle originally: If you make the measurement on any object, and you can determine the x-component of its momentum with an uncertainty Δp, you cannot, at the same time, know its x-position more accurately than $\Delta x = h/\Delta p$. The uncertainties in the position and momentum at any instant must have their product greater than Planck's constant. This is a special case of the uncertainty principle that was stated above more generally. The more general statement was that one cannot design equipment in any way to determine which of two alternatives is taken, without, at the same time, destroying the pattern of interference.

Let us show for one particular case that the kind of relation given by Heisenberg must be true in order to keep from getting into trouble. We imagine a modification of the experiment of Fig. 6–3, in which the wall with the holes consists of a plate mounted on rollers so that it can move freely up and down (in the x-direction), as shown in Fig. 6–6. By watching the motion of the plate carefully we can try to tell which hole an electron goes through. Imagine what happens when the detector is placed at $x = 0$. We would expect that an electron which passes through hole 1 must be deflected downward by the plate to reach the detector. Since the vertical component of the electron momentum is changed, the plate must recoil with an equal momentum in the opposite direction. The plate will get an upward kick. If the electron goes through the lower hole, the plate should feel a downward kick. It is clear that for every position of the detector, the momentum received by the plate will have a

Quantum Behavior

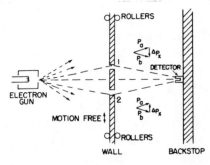

Figure 6-6. An experiment in which the recoil of the wall is measured.

different value for a traversal via hole 1 than for a traversal via hole 2. So! Without disturbing the electrons *at all*, but just by watching the *plate*, we can tell which path the electron used.

Now in order to do this it is necessary to know what the momentum of the screen is, before the electron goes through. So when we measure the momentum after the electron goes by, we can figure out how much the plate's momentum has changed. But remember, according to the uncertainty principle we cannot at the same time know the position of the plate with an arbitrary accuracy. But if we do not know exactly *where* the plate is we cannot say precisely where the two holes are. They will be in a different place for every electron that goes through. This means that the center of our interference pattern will have a different location for each electron. The wiggles of the interference pattern will be smeared out. We shall show quantitatively in the next chapter that if we determine the momentum of the plate sufficiently accurately to determine from the recoil measurement which hole was used, then the uncertainty in the x-position of the plate will, according to the uncertainty principle, be enough to shift the pattern observed at the detector up and down in the x-direction about the distance from a maximum to its nearest minimum. Such a random shift is just enough to smear out the pattern so that no interference is observed.

The uncertainty principle "protects" quantum mechanics. Heisenberg recognized that if it were possible to measure the momentum and the position simultaneously with a greater accuracy, the quantum mechanics would collapse. So he proposed that it must be impossible. Then people sat down and tried to figure out ways of doing it, and nobody could figure out a way to measure the position and the momentum of anything—a screen, an electron, a billiard ball, anything—with any greater accuracy. Quantum mechanics maintains its perilous but accurate existence.

Index

Index

About Richard Feynman

Born in 1918 in Brooklyn, Richard P. Feynman received his Ph.D. from Princeton in 1942. Despite his youth, he played an important part in the Manhattan Project at Los Alamos during World War II. Subsequently, he taught at Cornell and at the California Institute of Technology. In 1965 he received the Nobel Prize in Physics, along with Sin-Itero Tomanaga and Julian Schwinger, for his work in quantum electrodynamics.

Dr. Feynman won his Nobel Prize for successfully resolving problems with the theory of quantum electrodynamics. He also created a mathematical theory that accounts for the phenomenon of superfluidity in liquid helium. Thereafter, with Murray Gell-Mann, he did fundamental work in the area of weak interactions such as beta decay. In later years Feynman played a key role in the development of quark theory by putting forward his parton model of high energy proton collision processes.

Beyond these achievements, Dr. Feynman introduced basic new computational techniques and notations into physics, above all, the ubiquitous Feynman diagrams that, perhaps more than any other formalism in recent scientific history, have changed the way in which basic physical processes are conceptualized and calculated.

Feynman was a remarkably effective educator. Of all his numerous awards, he was especially proud of the Oersted Medal for Teaching which he won in 1972. *The Feynman Lectures on Physics*, originally published in 1963, were described by a reviewer in *Scientific American* as "tough, but nourishing and full of flavor. After 25 years it is *the* guide for teachers and for the best of beginning

About Richard Feynman

students." In order to increase the understanding of physics among the lay public, Dr. Feynman wrote *The Character of Physical Law* and *Q.E.D.: The Strange Theory of Light and Matter*. He also authored a number of advanced publications that have become classic references and textbooks for researchers and students.

Richard Feynman was a constructive public man. His work on the Challenger commission is well known, especially his famous demonstration of the susceptibility of the O-rings to cold, an elegant experiment which required nothing more than a glass of ice water. Less well known were Dr. Feynman's efforts on the California State Curriculum Committee in the 1960's where he protested the mediocrity of textbooks.

A recital of Richard Feynman's myriad scientific and educational accomplishments cannot adequately capture the essence of the man. As any reader of even his most technical publications knows, Feynman's lively and multi-sided personality shines through all his work. Besides being a physicist, he was at various times a repairer of radios, a picker of locks, an artist, a dancer, a bongo player, and even a decipherer of Mayan Hieroglyphics. Perpetually curious about his world, he was an exemplary empiricist.

Richard Feynman died on February 15, 1988, in Los Angeles.